"The two words from which we get techno_ _, _ _ _ _ _ _ _ _ _ used by New Testament writers to describe Jesus. Technology and Jesus—and therefore technology and humanity—are inseparable and delicately linked. Technology, like the human body itself, is a good servant but a bad master. Craig Gay has written a learned and lucid reflection on how it can both help and hinder human flourishing."
John Ortberg, senior pastor of Menlo Church, author of *I'd Like You More If You Were More Like Me*

"One of the most critical conversations of our day is quite simply this: How do we manage the machines and technologies that intersect our lives in a way that is consistent with our core Christian commitments? Craig Gay in this volume makes an invaluable and essential contribution, helping his readers think critically and more clearly about aspects of our daily experience that we all too easily take for granted. And part of the strength of this contribution is that Gay insists we need to think theologically about technology—that is, to view technology and respond to technology in light of the Triune God and a biblical understanding of what it means to be the church. And, of course, to then respond to the challenge of our day in a way that is intentional, discerning, and hopeful."
Gordon T. Smith, president of Ambrose University, Calgary, Alberta

"Craig Gay is neither a luddite nor a technophile. He's something much different—a thoughtful, engaged theologian grappling with some of the core questions raised by modern technology and its apparent move toward disembodied life. Like Ellul and Postman before him, Gay notes many of the ills of technological society, but he moves beyond these seminal voices by offering a constructive rather than merely descriptive account. Even for someone like myself who tends to be a bit more optimistic regarding the theological potential of modern technology, Gay offers a helpful corrective that is as incarnational as it is hopeful. I recommend it."
Kutter Callaway, assistant professor of theology and culture at Fuller Theological Seminary, author of *Watching TV Religiously* and *Breaking the Marriage Idol*

"In his new and captivating book, Craig Gay issues a call to right stewardship of modern technological development, so that we might live into the fullness of our embodied, ordinary, created human being. The chapters of this penetrating cultural appraisal constitute a tour de force of social philosophy, economic history, and theology, equipping us to live more Christianly. We are given reason and hope to practice resurrection."
Susan S. Phillips, executive director and professor at New College Berkeley, author of *The Cultivated Life: From Ceaseless Striving to Receiving Joy*

"This superb book is crucially important in four ways. First, it provides a lucid and chilling overview of what we all know in our bones but find it hard to talk about coherently: the more that technology—especially automation, our devices, and the internet—makes our life easier, the less that increasingly disembodied life seems to flourish. Second, it shows how, through a series of well-meaning mistakes (the ways we have shaped our science, our religion, and our commerce, and then let them reshape us), we got ourselves into a fix. Third and most important, it outlines how a forgetfulness of the beliefs that shaped our culture—creation, incarnation, redemption—have led to our current problems with disembodiment and psychic homelessness. As Gay puts it, 'When the trajectory of modern technological development is away from ordinary embodied existence it is at odds with God's purposes for his world.' But fourth—and here this book departs from most other laments about modern technology—the book ends with a robust and detailed survey of some of the ways (from how we eat to what and how we worship) that we can remember what our skewed technologies have dis-membered. Everybody who wants to recover their full humanity should read this book."

Loren Wilkinson, emeritus professor of philosophy and interdisciplinary studies at Regent College

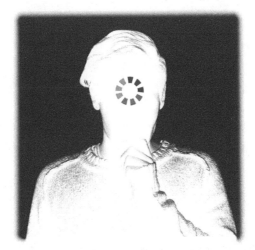

MODERN TECHNOLOGY AND THE HUMAN FUTURE

A CHRISTIAN APPRAISAL

CRAIG M. GAY

IVP Academic

An imprint of InterVarsity Press
Downers Grove, Illinois

InterVarsity Press
P.O. Box 1400, Downers Grove, IL 60515-1426
ivpress.com
email@ivpress.com

InterVarsity Press® is the book-publishing division of InterVarsity Christian Fellowship/USA®, a movement of students and faculty active on campus at hundreds of universities, colleges, and schools of nursing in the United States of America, and a member movement of the International Fellowship of Evangelical Students. For information about local and regional activities, visit intervarsity.org.

All Scripture quotations, unless otherwise indicated, are taken from The Holy Bible, New International Version®, NIV®. Copyright © 1973, 1978, 1984, 2011 by Biblica, Inc.™ Used by permission of Zondervan. All rights reserved worldwide. www.zondervan.com. The "NIV" and "New International Version" are trademarks registered in the United States Patent and Trademark Office by Biblica, Inc.™

Cover design: David Fassett
Interior design: Daniel van Loon
Images: © Andrea Leitgeb / EyeEm / Getty Images

ISBN 978-0-8308-5220-8 (print)
ISBN 978-0-8308-7384-5 (digital)

Printed in the United States of America ∞

InterVarsity Press is committed to ecological stewardship and to the conservation of natural resources in all our operations. This book was printed using sustainably sourced paper.

Library of Congress Cataloging-in-Publication Data
Names: Gay, Craig M., author.
Title: Modern technology and the Human Future: A Christian appraisal / Craig M. Gay.
Description: Downers Grove : InterVarsity Press, 2018. | Includes index. | Identifiers: LCCN 2018038710 (print) | LCCN 2018046469 (ebook) | ISBN 9780830873845 (eBook) | ISBN 9780830852208 (pbk. : alk. paper)
Subjects: LCSH: Theological anthropology--Christianity. | Technology--Religious aspects--Christianity. | Incarnation. | Forecasting.
Classification: LCC BT702 (ebook) | LCC BT702 .G39 2018 (print) | DDC 261.5/6--dc23
LC record available at https://lccn.loc.gov/2018038710

| P | 23 | 22 | 21 | 20 | 19 | 18 | 17 | 16 | 15 | 14 | 13 | 12 | 11 | 10 | 9 | 8 | 7 | 6 | 5 | 4 | 3 | 2 | 1 |
| Y | 38 | 37 | 36 | 35 | 34 | 33 | 32 | 31 | 30 | 29 | 28 | 27 | 26 | 25 | 24 | 23 | 22 | 21 | 20 | 19 | 18 |

For Andrew, Casey, Elsa, Owen, and Nicholas

When God lets himself be born and become [a human being], this is not an idle caprice, some fancy he hits upon just to be doing something, perhaps to put an end to the boredom that has brashly been said must be involved in being God—it is not in order to have an adventure. No, when God does this, then this fact is the earnestness of existence.

SØREN KIERKEGAARD, *THE SICKNESS UNTO DEATH*

CONTENTS

PREFACE

ON OCTOBER 26, 2016 (which happened to be my birthday) IBM ran a full two-page advertisement in the *New York Times* for its artificially intelligent super-computing system "Watson." "With Watson," the firm announced, "the world is healthier, more productive, more personal, more secure, more efficient, more creative, [and] more engaging." "Watson," the ad continued, "learns from us and extends our talents, augmenting our intelligence so that we can do our jobs better." Watson, we are told, will help us to "Do more. Do different. Do better."

As you read on, you'll see that I'm more than a little skeptical of these sorts of claims. Here at the outset I just want to make it clear that I've come by my skepticism "honestly," as they say. It also seems fitting, at the beginning of a book that will voice concerns about the depersonalizing impact of modern technologies, to start things off on a personal note.

I grew up in one of the mid-peninsula suburbs of San Francisco, California, that would become the hub of what today is called "Silicon Valley." My father was the Branch Manager for the San Francisco office of IBM, having worked for the company since the early 1950s. In 1970 my father left IBM to start his own computer-services company in—I kid you not—our garage, a venture that failed fairly quickly. The logo for that first enterprise was tacked up to the wall in our garage for decades thereafter. My dad's second venture, which he would also launch in our garage in partnership with two friends—who happened to be brilliant, but newly unemployed electrical engineers—would go on to prosper, becoming one of the early success stories of that era in "the Valley's" growth. The computer system the three partners developed was, if I'm not mistaken, one of the first to combine a micro-processing "chip" with a disk storage device. An exceedingly simple system by today's standards, it was basically a sophisticated (and rather large and expensive) "bean counter." Yet, it held revolutionary promise for several industries that had been struggling for some time to establish control over inventories of tens of thousands of individual parts. Beginning at the retail end, *Triad Systems Inc.* designed machines and procedures that eventually encompassed the entire supply chain, pioneering a "vertically integrated" marketing approach that was novel at that time.

I worked for *Triad* off-and-on for ten years starting at age fourteen or fifteen. I was the firm's first janitor. I also worked in Inventory Control and in Shipping and Receiving. I trained several generations of new salespeople in the use of the system and wrote user manuals for it. My first adult job was as a marketing applications representative, which meant coaching customers in the best use of new systems and/or new system features. I watched the firm grow from just a handful of employees into a $100-million corporation over those ten years. It was exhilarating. There was a real energy in the company, a sense that there were no limits to what was possible. I might have stayed with the firm had I not decided instead to go to graduate school.

My father and his partners took *Triad* public just after I'd finished university, splitting the proceeds with a venture capital firm that had backed them in the early days and that would go on to invest in Microsoft and other well-known technology firms. The IPO was modest by today's standards, but my father and his partners did well, realizing more than enough to live comfortably into retirement. I sold the shares that I had been given to pay for graduate studies.

I have certainly seen the upside of modern technological development. I've witnessed technology being created, and I benefitted very directly from it. Yet I've also witnessed its downsides. At the same time that my father's company was growing year on year, my wife's family's publishing company— located just a mile or two away—was experiencing the disrupting effects of new technologies, particularly the impact that the Internet began to have upon print advertising in the early 1990s. Reflecting back on the period, it's somewhat ironic, for my wife's father was very good friends with both Bill Hewlett and David Packard. *Sunset Publishing* had long been one of the anchors of our community, and while the business environment for magazines and books did not change overnight, print-advertising revenue and readership gradually began to dry up. The company's employees, meanwhile, found themselves increasingly priced out of a real estate market inundated by "tech money." Sunset relocated a number of years ago and, as of this writing, is up for sale, a victim apparently of the process of "creative destruction" that we will describe in due course. Modern technological development, I've observed, is good news for some but not particularly good news for others.

I must also confess to being an early adopter of technological gadgets. I finished my master's degree in 1982 on a Kaypro II, one of the first personal computers. I wrote my doctoral dissertation several years later on one of the early Macintosh computers (a 128K), a machine I still possess and that I believe still functions. I'm an Apple aficionado. I've had any number of iPods, iPhones, iPads, and so on. I've written most of this book on a MacBook Pro. I was an early convert to digital photography. I use a Garmin GPS-enabled cycling computer. I have a 3D TV (which I confess I only ever watch in 2D). Recently I've been experimenting with Google Home. And although I didn't ever actually build a Heathkit kit, I love going to Frye's Electronics whenever I'm back in my old hometown.

My wife and I have also raised four children since 1990, and we have witnessed—and been largely bewildered by—the impact of a raft of consumer technologies—including Game Boys, Play Stations, X-Boxes, the Internet, social media—upon our kids. Call me old fashioned, but while our kids are vastly more adept at using various technologies than I am (I still cannot type with my thumbs), I am waiting to be convinced that their lives have been significantly enriched by their use of these services and/or devices. In certain respects I think their lives have been impoverished by their use of them, and not simply because the services and devices have seemed only to get more and more expensive.

Now, my purpose in recounting all of this is to convince you that I am not a Luddite. I am not anti-technology. But I am concerned about it, and particularly about the directions it appears to be heading. I voice my concern not as technologist, not as an engineer, not even a businessman, but rather as a Professor of Interdisciplinary Studies at a small graduate school of Christian theology in Vancouver, Canada. Yet as the ancient Egyptian King Thamus is said by Socrates to have observed, "To one it is given to create the things of art, and to another to judge what measure of harm and of profit they have for those that employ them" (*Phaedrus* 274b-275c). It has been given to me—*I hope!*—to be of the latter sort, and my sincere desire is that you will find this book useful in judging what "measure of harm and of profit" you stand to gain by employing modern automatic machine technologies.

ACKNOWLEDGMENTS

THERE ARE, OF COURSE, MANY PEOPLE I need to thank for helping to make this book possible. The book's argument took shape over a number of years in a course entitled Christianity and Modern Technology, and I'm thankful for Don Lewis, Paul Williams, Jeff Greenman, and other deans who have enabled and encouraged me to teach it, as well as for all of the students who have taken it. I am grateful also to the Board of Governors and the administration of Regent College for approving the sabbatical leave during which most of the book was written. I am indebted to my faculty colleagues for their very helpful comments and suggestions following two faculty fora at which I presented early drafts of two chapters of the book. My good friends Ron Rittgers and Iain Provan read and commented helpfully upon the entire manuscript, and I am *very* grateful to and for them both. My friend Brian Williams invited me to a conference at Oxford University where I was able to run a précis of the book past a learned audience (that happened to include the Archbishop of Canterbury!). I've enjoyed the excellent research assistance of Ben Nelson, Kelsey May, and Kasey Kimball as well as the first-rate suggestions and copyediting of my friend Kathy Tyers Gillam. I am indebted also to David McNutt, associate editor for IVP Academic, for shepherding my manuscript through a remarkably timely and straightforward approval process. My intellectual debt to the late Peter L. Berger (Professor Emeritus of Religion, Sociology, and Theology at Boston University) is clearly evident in what follows. I am also deeply indebted to my friend, teacher, and mentor James M. Houston, who has inspired a great many of us to rediscover the profoundly personalizing implications of Trinitarian orthodoxy, and who encouraged me to read, and to *re*-read Jacques Ellul. Finally, I could not have written this book without the grace, love, and encouragement of my wife, Julie. *Thank you!*

INTRODUCTION

The very substance of our existing which has made us the leaders in technique, stands as a barrier to any thinking which might be able to comprehend technique from beyond its own dynamism.

GEORGE GRANT, "IN DEFENSE OF NORTH AMERICA"

WE LIVE IN EXTRAORDINARY TIMES, and we are beneficiaries of the astonishing increase in human productivity made possible by modern technology. It goes without saying that we should be deeply grateful for this. Not one of us would benefit from the quality of life we enjoy today if not for a series of technological revolutions that began toward the end of the eighteenth century and have continued with increasing frequency into the present—revolutions initially in the generation of an immense amount of useful mechanical energy, and now by means of a variety of digital technologies in the amplification of human mental energies.

Of course, we should be just as grateful for premodern technologies. I wouldn't have written this book, and you wouldn't be reading it, were it not for the astonishing genius of alphabetic literacy, typography, and book-making, all of which have profoundly transformed human consciousness—and on top of which digital word-processing is merely a footnote.

Technology is apparently something we human beings do, something that we develop more or less naturally. Yet our technologies move us beyond mere nature. All animals labor, Abraham Lincoln is said to have observed, but only human beings *improve* their workmanship by way of discoveries and innovations.[1] And our workmanship is not the only thing we improve by means of technology. Our discoveries and innovations have led to improvements, literally, in *ourselves*. For example, there is evidence that our cognitive capacities—so much greater and more flexible than those of our nearest evolutionary relatives—emerged only in conjunction with technological developments such as language acquisition, cooking, and tool use. It seems that the uniquely human way of being-in-the-world is indissolubly linked with technological making. Perhaps, as technology critic Nicholas Carr recently opined, the human will-to-technology arises out of that frustrating tension we so often experience between what our minds can envision and what our unaided bodies can accomplish.[2] "Through our tools," Carr suggests, "we give our dreams form. We bring them into the world. The practicality of technology may distinguish it from art, but both spring from a similar, distinctly human yearning."[3]

Yet while the will-to-technology may be a necessarily self-creative act, from time to time it oversteps its rightful boundaries. Modern technologies, for example, have unquestionably increased productivity, making new material freedoms possible, but they have also seemed to undermine the prudence we require to exercise these freedoms wisely and judiciously. Just as we have recently brought the goods of electricity and penicillin into the world, so we have also brought the evils of nuclear weapons and weaponized anthrax. Just as we once invented alphabetic literacy, so we have now invented all manner of technological gadgets that distract us from reading and even, more often than we might care to admit, from thinking. While our technologies can apparently empower us to become *more* of ourselves, they can also permit us to become *less*, diminishing us even as they purport to deliver "more" and "better," "faster" and "easier." A large and growing body of evidence suggests that the impact of modern technology—in particular, the impact of automatic machine technology—upon us is not altogether beneficial. The trajectory of modern machine development appears now to

---

[1]Abraham Lincoln, cited in William Rosen, *The Most Powerful Idea in the World: A Story of Steam, Industry, and Invention* (New York: Random House, 2010), 323.
[2]Nicholas Carr, *The Glass Cage: Automation and Us* (New York: W. W. Norton, 2014), 215.
[3]Ibid.

be diverging away from and not toward the enrichment of ordinary embodied human being.

This odd divergence is the problem that I plan to take up in these pages. Gratefully acknowledging modern technology's benefits and celebrating technological making as an essentially human enterprise do not obviate the need to think carefully about it, to evaluate it, and to try, where possible, to improve it. It is hoped that this book will encourage readers to better understand modern technological development in light of its economic context and distinctive history but also to begin to think about it from the standpoint of the impact that it is having upon ordinary embodied human being. My concern is that we are allowing ourselves to be diminished by our own technologies. This, I will argue, is something that we should resist.

To begin, let me clarify how I plan to use the terms "technology" and "modern technological development." By technology I mean the systematic application of knowledge, methods, and tools to various practical tasks. I have in mind the various devices and systems that we employ and are employed by and that ordinarily surround us today, from iPods and laptop computers to medical information systems and Amazon.com. By modern technological development, I mean the direction(s) that our devices and systems appear to be heading and what their continued growth might portend for us. Our technologies disclose and shape our self-understanding and our way of being-in-the-world. As mentioned above, they can enable us to become more of ourselves, rendering our relations with each other and with the world richer and more meaningful. Yet our technologies can also inadvertently diminish us. We can find ourselves overcome and mastered by the very tools we have developed and deployed to provide us with mastery. Just how and why modern automatic machine technologies have tended toward this latter outcome is something we will explore.

When it comes to evaluating modern technological development, there have long been concerns that modern technology is somehow at odds with human flourishing. From William Blake's characterization of early industrial technology in terms of "dark Satanic Mills," to Mary Shelley's chilling account of Dr. Frankenstein's trans-human monster, to Martin Heidegger's fear that modern technology obscures the human *essence*, to recent concerns about

widespread and permanent computer-driven "technological unemployment"[4]
—any number of observers have described something *in*human and ultimately
*de*humanizing about modern machine technology. "I have a plain sense," the
philosopher and Roman Catholic thinker Romano Guardini wrote, reflecting
upon the impact of modern technology on European culture (ca. 1923), "that
a world is developing in which human beings . . . can no longer live—a world
that is in some way nonhuman."[5] Or, as the Russian Orthodox thinker Nicholas
Berdyaev noted somewhat more explosively (ca. 1935), "The chief cosmic force
which is now at work to change the whole face of the earth and dehumanize
and depersonalize man is . . . technics, the wonder of our age. Man has become
a slave to his own marvelous invention, the machine."[6] Or as the American
historian of technology, Lewis Mumford, posed—presciently—in 1964,

> What merit is there in an over-developed technology which isolates the whole
> man from the work-process, reducing him to a cunning hand, a load-bearing
> back, or a magnifying eye, and then finally excluding him altogether from the
> process unless he is one of the experts who designs and assembles or pro-
> grams the automatic machine? What meaning has a man's life as a worker if
> he ends up as a cheap servo-mechanism, trained solely to report defects or
> correct failures in a mechanism otherwise superior to him? If the first step in
> mechanization five thousand years ago was to reduce the worker to a docile
> and obedient drudge, the final stage automation promises today is to create a
> self-sufficient mechanical electronic complex that has no need even for such
> servile nonentities.[7]

British journalist Bryan Appleyard recently voiced similar concerns: "It
does not matter whether the new machines never achieve full human-like
consciousness, or even real intelligence, they can almost certainly achieve
just enough to do your job—not as well as you perhaps, but much, much

---

[4]A term coined ca. 1930 by John Maynard Keynes but employed most recently by Erik Brynjolfsson and Andrew McAfee in *The Second Machine Age: Work, Progress, and Prosperity in a Time of
Brilliant Technologies* (New York: W. W. Norton, 2014); see also Jaron Lanier, *Who Owns the Future?* (New York: Simon & Schuster, 2013).

[5]Romano Guardini, *Letters from Lake Como: Explorations in Technology and the Human Race*
(Grand Rapids: Eerdmans, 1994 [1923]), 7.

[6]Nicholas Berdyaev, *The Fate of Man in the Modern World* (Ann Arbor: University of Michigan
Press, 1935), 80-81.

[7]Lewis Mumford, *The Myth of the Machine: The Pentagon of Power* (New York: Harcourt Brace
Jovanovich, 1964), 179.

more cheaply."[8] In a modern world in which more and more decisions boil down to monetary calculations, it is not terribly surprising that automated systems have increasingly impacted the lives of ordinary people. "To modernise John Ruskin," Appleyard writes, "there is hardly anything in the world that some robot cannot make a little worse and sell a little cheaper, and the people who consider price only are this robot's lawful prey."[9] Late educator and critic Neil Postman summarized these sorts of concerns about modern machine technology in 1992 in the neologism *technopoly*, that sorry state into which modern American civilization has already fallen and in which everything that passes for culture—religion, art, philosophy, morality, law, politics—has been surrendered to the logic of technology.[10]

Trans-humanizing or De-humanizing?

There are those today who see technological dehumanization as a kind of necessary and inevitable aspect of human evolution, as if to say, "Ordinary embodied humanity? Good riddance! 'Humanity' as we presently understand the notion is overrated and fleeting." Modern digital technology holds out the promise—conceivably realizable in the very near future—of an entirely new way of being, a way of being more than merely human. Indeed, digital technology holds out the promise of a *trans*-human existence, free from the physical limitations of time and space, even free ultimately from death.

My hunch is that most of us, while we are undoubtedly grateful for advances in medical technology, as well as for the efficiencies and conveniences offered by computerized devices, do not actually long for *trans*-human immortality. Rather, we are simply hoping for more life now, and we harbor the suspicion that our technologies are somehow getting in the way of it. Such was the thrust of American philosopher Albert Borgmann's argument in *Technology and the Character of Contemporary Life*.[11] We are drawn to

---

[8]Bryan Appleyard, "The New Luddites: Why Former Digital Prophets Are Turning Against Tech," *New Statesman*, August 29, 2014, www.newstatesman.com/sci-tech/2014/08/new-luddites-why-former-digital-prophets-are-turning-against-tech.
[9]Ibid.
[10]Neil Postman, *Technopoly: The Surrender of Culture to Technology* (New York: Vintage, 1993), 71.
[11]Albert Borgmann, *Technology and the Character of Contemporary Life: A Philosophical Inquiry* (Chicago: University of Chicago Press, 1984).

modern technological devices, Borgmann argued, because they promise to make an expanded range of things and possibilities conveniently available to us without burdening us with the need to know how they work—or how, much beyond turning them on and off, to operate them. And yet as more and more of our time is spent consuming the commodities that our technological devices procure for us, we lose touch with the skill development and habits and the social environments that traditional (which is to say "pretechnological" and today is often read as "outdated") practices used to require and that used to, as Borgmann puts it, "grace and orient our lives."[12]

Borgmann illustrates this tradeoff with a telling comparison between giving a child a musical instrument, with lessons, and giving him or her a stereo set (he was writing, of course, before the appearance of the iPod). Both gifts have to do with making music. Indeed, the latter produces professional-quality music "right out of the box," as they say. Still, the former gift—because it requires discipline and patience, as musical skills are painstakingly developed over time—stands a chance of making a significant and lasting impact upon the child's growth and development, instilling habits and producing character, as well as perhaps introducing him or her to a life-long avocation.

Yet in spite of the potential benefits of musical education, most of us today opt instead to purchase digital music players and/or subscriptions to streaming services (or iPods) for our children, for these are what our children tell us that they want. And so we are deceived by the cleverness and commercial availability of an ever-expanding array of systems and devices into neglecting the very things and practices that once added depth to our experience of the world and which made our lives worth living, as the saying goes. We are subsequently surprised to discover, after all of the things that used to "burden" us have been conveniently removed, that our lives have become disengaged, distracted, and lonely. But we shouldn't be surprised. It is fairly obvious, Borgmann concludes, that "technological liberation from the duress of daily life is only leading to more disengagement from skilled and bodily commerce with reality."[13]

---

[12]Ibid., 157.
[13]Ibid., 151.

So what do we make of this inadvertent disengagement? What, beyond trying to turn back the clock, can we possibly do about it? One answer is simply to accept it and to hope that we will, somehow, successfully adapt to it. This was the gist of sociologist Francis Fukuyama's analysis of the technologically driven change that we have recently experienced in transitioning from an industrial to a *post*-industrial social environment, a study aptly titled *The Great Disruption: Human Nature and the Reconstitution of Social Order*.[14] "Broadly speaking," Fukuyama opined, "the technological change that brings about what economist Joseph Schumpeter called 'creative destruction' in the marketplace [has] caused similar disruption in the world of social relationships [i.e., evidenced by criminality, divorce and familial disintegration, substance abuse, mental illness, declining life-satisfaction, suicide, etc.]. It would be surprising were this not true."[15] The upside, as the book's subtitle suggests, is that human beings are intrinsically social animals who will somehow and eventually figure out how to reconstruct life satisfaction amidst their new circumstances. There is hope, Fukuyama concluded, because of "the very powerful innate human capacities for reconstituting social order."[16]

More recently and along a similar line, technology journalist Clive Thompson has argued in *Smarter Than You Think: How Technology Is Changing Our Minds for the Better* that our minds have always worked in tandem with our tools, and that as our tools have evolved, so inevitably have we.[17] Indeed, on the basis of what is called the "extended mind theory" of cognition, Thompson argues that the very reason the human species is so intellectually dominant is because we've managed to "outsource" aspects of cognition to technological systems—from clay tablets, to pens and ink, to RAM—in effect scaffolding our thinking into newer and more creative realms. The powerful digital tools now at our disposal simply mean that we can look forward to becoming even smarter. Thompson writes,

---

[14]Francis Fukuyama, *The Great Disruption: Human Nature and the Reconstitution of Social Order* (New York: Free Press, 1999).

[15]Ibid., 6.

[16]Ibid., 282.

[17]Clive Thompson, *Smarter Than You Think: How Technology Is Changing Our Minds for the Better* (New York: Penguin, 2013).

Search engines answer our most obscure questions; status updates give us an ESP-like awareness of those around us; online collaborations let far-flung collaborators tackle problems too tangled for any individual. We're becoming less like Rodin's *Thinker* and more like . . . centaurs. This transformation is rippling through every part of our cognition—how we learn, how we remember, and how we act upon that knowledge emotionally, intellectually, and politically. . . . [Our] tools can make even the amateurs among us radically smarter than we'd be on our own.[18]

Unfortunately, there isn't much evidence that any of this is actually happening. The evidence thus far suggests that while technology may well be changing our minds, it is not changing them for the better. Google may not actually be making us stupid, but it has not (yet?) made us much smarter.

Of course, we will adapt. What choice do we have? Yet the suggestion that we will eventually adapt to our technologies is curious. Why should we have to? Aren't our technologies supposed to make our lives better and easier? Aren't our tools supposed to serve us? Yet one would be hard pressed to say that human beings equipped with the latest technologies are happier, more contented, more fulfilled, and better-adjusted to their environments than those not similarly outfitted.

And this isn't a particularly new problem. As Lewis Mumford quipped nearly a hundred years ago in his celebrated history of technological development, *Technics and Civilization*, a good deal of our technical apparatus is useful in the same way that crutches are useful when one's leg is broken: they're better than nothing, but it's far easier to get around on normally functioning legs. "The common mistake," Mumford continued, "is that of fancying that a society in which everyone is equipped with crutches is thereby more efficient than one in which the majority of people walk on two legs."[19] G. K. Chesterton contended that this common mistake disclosed what he called "the huge modern heresy," a kind of civilizational error in which it comes to be believed that the human soul must somehow be made to fit the requirements of modern technological systems, rather than ordering these systems to the requirements of the human soul.[20]

---

[18]Ibid.
[19]Lewis Mumford, *Technics and Civilization* (London: Routledge & Sons, 1934), 276.
[20]G. K. Chesterton, *What's Wrong with the World?* (London: Cassell & Co., 1910), 109.

One of the central contentions of this book is that Chesterton was right, and that requiring human persons to adapt to the impersonal workings and rhythms of modern machine systems is gravely mistaken. We need instead to reorder our technologies to the needs and requirements of ordinary embodied human persons. We need to adjust our technological systems such that they enable us to become *more*, and not less, of ourselves.

The issue of the relation of our technologies to human nature and/or to the human soul has been hotly debated within Western civilization, often making reference to Socrates's hapless lament that encouraging literacy would only undermine human memory.[21] And the issue will surely continue to be debated. In the meantime, technological possibilities are evolving rapidly. As computer security expert Thomas Keenan recently warned in a book provocatively titled *Technocreep: The Surrender of Privacy and the Capitalization of Intimacy*,[22] there is actually much more to modern digital technology than meets the eye. Systems that most of us are not even aware of are even now operating continuously in the background of our lives, systems that are increasingly beyond anyone's direct control. "Like a network of mushroom spores sending out subterranean tendrils to silently exchange genetic material," Keenan writes, "our technological systems are increasingly passing information back and forth without bothering to tell us. They are parsing and analyzing it to squeeze out the deep meaning of what we say and do, sometimes before we are even aware of our own intentions."[23]

Personal privacy is not the only issue at stake. As Carr observes in his recent study of automation, this subterranean infrastructure of automated systems is likely to account for the lion's share of "technological unemployment," possibly in the near future.[24] These systems have been expressly designed and developed, not simply to be unobtrusive and invisible, but so as not to require human oversight. We would do well to be concerned about

---

[21]Plato, "Phaedrus," in *The Collected Dialogues of Plato Including the Letters*, ed. Edith Hamilton and Huntington Cairns, Bollingen Series 71 (Princeton, NJ: Princeton University Press, 1964), 520.

[22]Thomas P. Keenan, *Technocreep: The Surrender of Privacy and the Capitalization of Intimacy* (Vancouver; Berkeley, CA: Greystone, 2014).

[23]Ibid., 11.

[24]Carr, *The Glass Cage*, 198.

this, Carr contends, for when inscrutable technological systems become invisible, the assumptions and intentions that have been built into them are likely to have so infiltrated our own desires and actions that it will be impossible to discern whether the algorithms are helping us or controlling us.[25]

Toward an Evaluation of Modern Technology

Clearly we have work to do, and soon. For even if, like Fukuyama and Thompson, we are optimistic about the malleability and adaptability of human nature in the face of new technologies, it's not clear how we could adapt to an interlocking array of technological systems that no longer require human involvement as anything but minions. The only way of avoiding such a fate, short of natural and/or geo-political disaster, may be in trying to elaborate a new way—or, as I plan to argue in the following pages, in reclaiming an extraordinarily fruitful *old* way—of evaluating and developing our tools and technologies.

In saying this, we need to be careful not to construe our present predicament in terms of problems for which solutions must somehow be devised. Such a construal—one toward which we are naturally, almost reflexively drawn—simply discloses the prevailing, largely technological ethos. Modernity, after all, is all about devising technical solutions for various problems. Indeed, as sociologist Peter Berger observed, modern societies are animated by the assumption that "all human problems can be converted into technical problems, and if the techniques to solve certain problems do not as yet exist, then they will have to be invented."[26] This approach quickly develops into a kind of practical anthropocentrism, in which we come to imagine the world as largely meaningless apart from those uses that we have determined to make of it.[27] Another word for such an outlook is *nihilism*, and a kind of implicit or soft nihilism pervades contemporary technological culture in the form of the increasingly common beliefs that it is up to us to "construct" ourselves and that truth is simply a kind of ideational means to the end of generating satisfying human purposes.

---

[25]Ibid., 210.

[26]Peter L. Berger, *Pyramids of Sacrifice: Political Ethics and Social Change* (Garden City, NY: Anchor, 1974), 20.

[27]Gabriel Marcel, "The Sacred in a Technological Age," *Theology Today* 19 (1962): 28-29.

Yet in addition to being rather exclusively focused on human needs and human desires, modern technological development often feels strangely alien and unassailable. Having become so remarkably adept at "inventing technical solutions to human problems," it seems we are now hard-pressed to imagine any other way of being-in-the-world, thus sanctioning the dense web of technical solutions in which we so often find ourselves entangled nowadays. Philosopher George Grant put our difficulty this way a number of years ago in the memorable passage cited at the outset of this chapter:

> We [in North America] live in the most realised technological society which has yet been; one which is, moreover, the chief imperial centre from which technique is spread around the world. . . . Yet the very substance of our existing which has made us the leaders in technique, stands as a barrier to any thinking which might be able to comprehend technique from beyond its own dynamism.[28]

Our inability to "comprehend technique from beyond its own dynamism" is one of the chief reasons we have frequently allowed modern technology to run roughshod over actual human beings. "That's just progress," we say, trusting that the benefits of newer and better technologies must always—somehow—outweigh their human costs. And perhaps in many respects they have. Yet we seem to have arrived at the point now where it's difficult to see what further benefits could justify modern technology's mounting human costs. What technological breakthrough could possibly justify the neglect and diminution of ordinary embodied human being?

Yet perhaps this is to say too much and too quickly. If we are to grasp the significance of this inability to "comprehend technique from beyond its own dynamism," we must at least suspect that modern machine technology actually poses a problem and we must at least have begun to become convinced that it does in fact threaten to diminish ordinary embodied human being. Convincing the reader of this is the purpose of our first chapter. The chapter begins by recalling the wonderfully constructive impact that several *pre*-modern technologies had upon human consciousness, but then goes on to catalog

---

[28]George Grant, "In Defense of North America," in *Technology and Empire: Perspectives on North America* (Toronto: Anansi, 1969), 40.

some of the recent evidence that suggests that the impact that modern automatic machine technology is having upon us is not altogether positive.

The accurate assessment of modern technological development, however, is often frustrated by powerful economic forces that extend beyond the interests of entrepreneurs, patent holders, and venture capitalists to include all who depend upon economic growth—which is to say, nearly all of us. Critics of modern technology often overlook the crucial connection between technological development and economic interests. It was capitalism that provided the context within which modern automatic machine technology was first developed. Not only did the market system create powerful incentives for making mechanical improvements, it also provided the capital for experimentation and then for bringing these improvements to market. Were it not for capitalism's direction of resources, the technological ethos might have remained merely academic and largely socially irrelevant. The modern market system has connected the new mechanical orientation to the lives and concerns of ordinary people, and it continues to interpret what we will call, in chapter two, the momentum and inertia of modern technological development.

Yet diagnosing our relative inability to gain a truly objective view of modern technology must be extended beyond economic forces—and even beyond our obvious skill at technological making. Our inability appears even more basically to stem from a kind of "mechanical world picture" within which nature—including human nature—is often conceived on the analogy of a vast and elaborate mechanism, one that differs from automatic machine technology in complexity and scale but not in kind. From within this world picture, ordinary embodied human existence is often construed not as something to be nurtured and enhanced, perhaps by technology, but rather as a series of limitations that remain to be overcome with more and better technology. This mechanical world picture has a long and interesting history within Western civilization that culminates in the distinctively modern "scientific" outlook. We will review this history in some detail in our third chapter, for it is a history that must be understood if we are to stand any chance of properly evaluating the disembodying thrust of modern technological development.

Our current inability to "comprehend technique from beyond its own dynamism" also helps to explain why even profoundly insightful criticism of modern technology has not tended to alter the course of its development. Neil Postman is rightly famous for his incisive criticism of modern educational and other technologies, criticism intended to expose the narrow and constrictive technical logic implicit in so many modern programs and systems. Yet can the reader recall the solution Postman posed to the technological absurdities he uncovered? It consisted in a rather insipid proposal to revive a kind of "Great Books" curriculum, in which subjects were to be presented as stages in "humanity's historical development," which would—somehow (?)—enable us to determine how best to direct further technological development.[29] Beyond alarming us, in other words, Postman's analyses didn't leave us much to work with.

More recent criticism of technology suffers from a similar handicap. Sherry Turkle's careful analysis and excellent critique of social networking and digital media and their impact upon young adults, *Alone Together: Why We Expect More from Technology and Less from Each Other*, also offers little in the way of a constructive response to deeply troubling problems.[30] At the end of the day, Turkle is hard pressed to say just what is wrong with the sentiment, expressed by one young woman, that she would probably prefer the companionship of a reliable robot to that of an actual—likely unpredictable—male human being. In short, recent criticism of technology tends to exemplify the mischievous quip, sometimes attributed to G. K. Chesterton, that modern critics are almost always right about what is wrong, and almost always wrong about what is right.

My hope is that this book will prove an exception to Chesterton's rule. In seeking not to be wrong about what is right, I want to heed another piece of Chesterton's perennially sage advice. When things are going wrong, he observed, you don't need practical solutions so much as you need philosophical commitments, firm convictions about the way things are—or at least about the way things ought to be. The strength to set things right never comes from

---

[29]Postman, *Technopoly*, 198-99.
[30]Sherry Turkle, *Alone Together: Why We Expect More from Technology and Less from Each Other* (New York: Basic Books, 2011).

criticism alone. Rather, it comes from convictions that are often religious and, therefore, are religiously held.[31] In this connection, I believe that the only framework that is up to the task of constructively criticizing the disembodying/dehumanizing thrust of modern technology—indeed the only framework that can defend embodied human existence under modern conditions—is derived from the central tenets of the *Christian* religion, specifically from the orthodox Christian doctrines of *creation, incarnation,* and *resurrection.* No other philosophy, ancient or modern, religious or secular— save perhaps Orthodox Judaism—comes close to valuing embodied human existence as highly as orthodox Christianity.

We will have much more to say about this in our fourth chapter, and we will explore the implications of Christianity's astonishingly high view of human embodiment in chapter five. Suffice it here at the outset to recall that the Nicene Creed declares—again, astonishingly—that Jesus Christ, the one who is "God from God, Light from Light, true God from true God, begotten, not made, of one Being with the Father, through whom all things were made" was *incarnate* and *made human.* The creed goes on to affirm that Jesus Christ was raised *bodily* from the dead, that he has ascended *bodily* into heaven, where he now "sits at the right hand of the Father," and in its Armenian version the creed concludes with the declaration that Jesus Christ will come again "*with the same body* . . . to judge the living and dead." It is simply not possible to dignify the human body any more highly than this.

From the point of view of the Christian religion, then, modern technology's diminishment of ordinary embodied human existence poses a very serious problem. If the Christian proclamation of the incarnation signals the divine intention to redeem, restore, and ultimately to glorify embodied human beings, then clearly anything that undermines, enfeebles, or otherwise diminishes ordinary embodied human being must be at odds with the divine purpose. We must decide, then—given its evident trajectory—whether modern automatic machine technology qualifies for this kind of censure.

On a more positive note, Christian theology affirms that our bodies have been designed for relationship with each other, with God, and with the rest

---

[31]Chesterton, *What's Wrong,* 43.

of created nature. This puts an entirely new spin on what we ought to use our technologies *for*. It suggests, for example, that our technologies ought to be used to enhance embodied relationality, not necessarily to escape from bodily limitations. It suggests that while we ought to develop and use our tools and technologies to enhance all aspects of personal relationality, we ought still especially to prioritize embodied, face-to-face relations.

But wait (the discerning reader will rightly object), is not "modernity" itself the product of Christian civilization? Isn't the Christian religion often credited with nurturing the intellectual seedbeds of modern science and technology? Indeed, isn't the Christian religion to blame for the phenomenon of disembodiment within Western society and culture? And even assuming that the Christian religion can somehow be exonerated of these charges, why has Christian theology not already critiqued and solved the applicable problems within modern society and culture?

These are legitimate questions. Two things may be said at the outset: The first is simply to confess that the implications of Christian theology for the human body have been largely forgotten, even by the Christian church, and I hope to shed some light on how and why. Second, one reason why Christian convictions about the human body's significance currently lack influence within contemporary society and culture is because both society and culture have increasingly become *post*-Christian. Having lost sight of the Christian hope of salvation for our world, modern men and women have resorted to a kind of *gnostic* hope of salvation from this world. We will say more about this, but the modern variation on the ancient theme of gnosticism is one in which nature—which includes the human body—is to be mastered by the human spirit (largely by means of technology) and altered as necessary, for the sake of purposes that human beings have willed. Up until fairly recently, gnostic themes were simply implicit within the modern technological worldview, but they have surfaced explicitly and with increasing frequency at the forefront of technological research and development. Indeed, "transhumanism"—the belief that human beings must evolve beyond their current embodied limitations—has increasingly become part of our cultural vernacular.

Modern Technology and the Human Future is obviously intended primarily for a Christian audience. I will ask the reader throughout to take seriously the core Christian claim that God has, in Christ, become an embodied human person. If this is true, then there can be no post- or transhumanism, rather, only technologies that either enhance or diminish what is genuinely human. The judgments we make along these lines will form the basis for determining whether we will use or not use, support or not support, various modern technologies.

Yet it is worth noting that an analogous—if much less robust—defense of human embodiment can be made on purely secular grounds. On the one hand, as Albert Borgmann argued in a series of lectures at Regent College in 2010, when technological development drifts away from embodied human being, it must inevitably dissolve into pointlessness and indeterminacy. "In the disembodied world of computerized hyperintelligence, the number and kinds of artificial experiences are boundless," Borgmann argued, yet "when a near infinity of the most diverse experiences are available to an [artificial intelligence], nothing in particular stands out and commands attention anymore."[32] When anything and everything is instantly possible and at every moment, why go to the trouble of actualizing anything?

If, furthermore, we believe that human programmers can somehow supply "purposes" to artificially intelligent machines, we are simply mistaken; humanly recognizable purposes only make sense in the context of embodied human being. The human mind didn't accidentally end up in the human brain, Borgmann reasoned, and the human brain doesn't just happen to be housed in a human body. Rather, all three—mind, brain, and body—have evolved together and are essentially one and interdependent.[33] Whatever else disembodied "intelligence" may be said to be, it will not—it *cannot*—ever be human. Borgmann's argument simply implies that we had better put our efforts into being what we actually are—embodied human beings in a

---

[32]Albert Borgmann, "Pointless Perfection and Blessed Burdens," *Crux*, 47, no. 4 (2011): 24; the mid-twentieth-century work of philosopher Maurice Merleau-Ponty is often cited in this connection; he insisted upon the inseparability of human consciousness from corporality. Human consciousness does not simply take place in the human body, rather it is inextricably bound up with the human body. "To understand," Merleau-Ponty writes, "is to experience harmony between what we aim at and what is given, between the intention and the performance—and the body is our anchorage in the world" (Maurice Merleau-Ponty, *Phenomenology of Perception*, trans. Colin Smith [London: Routledge & Kegan Paul, 1962], 144).

[33]Ibid., 23.

world delimited by time and space—rather than into trying to become something that we cannot ever be.

For similar reasons, it isn't difficult to see that the optimistic suggestion that human beings will quickly adapt to technological disembodiment must also be erroneous. Our minds, brains, and bodies are actually quite remarkably suited to the physical and social environments in which we still, for the most part, find ourselves. While it is true that we have managed to adapt—typically over many centuries—to using new tools and technologies, it amounts to a kind of cruel madness to imagine that in a matter of just a few decades we can engineer a fundamentally new and different way of being-in-the-world.

But enough said. Before suggesting solutions, it is first necessary to convince you, the reader, that there is a problem—that modern automatic machine technology is indeed diminishing us as persons in various ways. Examining evidence of this diminishment from a variety of sources sets the agenda for our first chapter.

CHAPTER ONE

MACHINE TECHNOLOGY
AND HUMAN BEING

[There] is a paradox common to technological existence:
everything gets a little easier and a little less real.

ALEX ROSS, "THE RECORD EFFECT"

AMIDST WIDESPREAD ENTHUSIASM for all the things that will soon be
made possible by newer and better technologies—from advanced genetically
engineered cancer treatments, to ultra-safe self-driving automobiles, to en-
thralling virtual environments—the assertion that modern technology is diminishing ordinary embodied human being is both countercultural and conten-
tious. Still, the purpose of this first chapter is to try to persuade even the skeptical
reader that, given its evident trajectory (Silicon Valley boosterism notwithstanding), modern automatic machine technology is more likely to detract from
our ordinary embodied experience of the world than it is to enhance it.

This should not be taken to mean that the problem lies with technology
per se. On the contrary, and as we suggested in the introduction, technological making is indissolubly linked with the distinctively human way of
being-in-the-world. Our technologies often enable us to become more of
ourselves, more personally interrelated with each other, more dynamically
engaged with the world, and better able, even, to worship the living God. So
it may help to preface our consideration of modern technology's impact of
diminishing human existence by considering the remarkably positive impact
that several premodern technologies had upon human consciousness, technologies that we take for granted when we read and write.

Writing, Reading, and Human Response-ability Before God

The Christian religion summarizes human purposes in the double commandment of love. We are commanded to love God with all of our heart, mind, soul, and strength; and we are commanded to love our neighbor as ourselves. For most of us, most of the time, both commands are only aspirational. It is precisely in considering the demands of the "Great Commandment" that we are most likely to be persuaded of our deep need for forgiveness. Love, after all, is an act of self-transcendence. It entails the giving of oneself to another. It must be freely offered. It is often costly. And love can only arise out of mature personal agency. In this connection, we commonly distinguish self-transcending love—*agapē*—from merely affective and/or erotic love. We also distinguish mature adult love from the love of children; a child's love is delightful, but it remains undeveloped and as yet unable give a full account of itself. We simply do not expect childish love to be wholly self-transcending.

While human beings have surely always been capable of loving friends and relatives, cultural historians have suggested that the responsible personal agency necessary for *agapē* is not a feature of "human nature" per se. Rather, it appears to have emerged culturally and historically. For responsible personal agency to have arisen, human beings apparently needed to be lifted out of an original and, as it were, childlike immersion in the immediacy of nature.[1] In the first volume of his remarkable historical study of social order, *Order and History*, entitled *Israel and Revelation*, Eric Voegelin described the original human experience as follows:

> Whatever man may be, he knows himself a part of being. The great stream of being in which he flows while it flows through him, is the same stream to which belongs everything else that drifts into his perspective. . . . We move in a charmed community where everything that meets us has force and will and feelings, where animals and plants can be men and gods, when men can be divine and gods are kings.[2]

---

[1] I have taken portions of the following argument concerning Voegelin and Israel's "leap in being" from my article "Christianity and the 'Homelessness' of the Modern Mind," *Christian Scholar's Review* 23, no. 2 (December 1993): 127-44.

[2] Eric Voegelin, *Israel and Revelation*, vol. 1 of *Order and History* (Baton Rouge: Louisiana State University Press, 1956), 3.

Voegelin's purpose in trying to represent this original experience for the modern reader was twofold: In the first place, he wanted to suggest that to know oneself as "a part of being" is not yet to know oneself as capable of responsible "historical" agency. Voegelin also wanted his modern readers to see that, for responsible historical agency to have become a possibility, human awareness needed somehow to be lifted out of the participatory worldview of archaic cosmology. Human beings needed to see that they could stand apart from and in a sense above the great stream of natural being.

Voegelin then went on to argue that human consciousness experienced a kind of "leap in being"[3] in ancient Israel's experience of revelation, that is, in the nation's founding declaration that a God who stood entirely outside and above nature had entered into the natural continuum and had spoken personally to Abraham, Isaac, Jacob, and eventually to Moses, calling them into relationship with himself. "Suppose . . . they ask me, 'What is his name?' Then what shall I tell them?" Moses inquired of the voice that had addressed him from out of the oddly—unnaturally—burning bush (Ex 3:13). "I AM WHO I AM," came the astounding reply. "Say to the Israelites, 'The LORD, the God of your fathers—the God of Abraham, Isaac, and Jacob—appeared to me and said: I have watched over you. . . . And I have promised to bring you up out of your misery in Egypt into a land flowing with milk and honey'" (Ex 3:15-17).

Israel's exodus from Egypt was, thus, far more than simply a demographic event. Indeed, as sociologist Peter Berger observed,

> It constituted a break with an entire universe. At the heart of the religion of ancient Israel lies the vehement repudiation of both the Egyptian and the Mesopotamian versions of cosmic order, a repudiation that was, of course, extended to the pre-Israelite indigenous culture of Syria-Palestine. The "fleshpots of Egypt," from which Yahweh led Israel into the desert, stood above all for the security of the cosmic [the natural] order in which Egyptian culture was rooted.[4]

What Yahweh's address to Israel called forth, in such a way as to permanently alter the course of human history, was the possibility of dialogue with

---

[3]Ibid., 130.
[4]Peter L. Berger, *The Sacred Canopy: Elements of a Sociological Theory of Religion* (Garden City, NY: Anchor, 1969), 115.

the living God and hence of a fundamentally new way of being-in-the-world, a way-of-being that would no longer place the primary emphasis upon the maintenance of cosmic order, but would insist instead upon the paramount importance of individual, personal, responsible, and historical human agency before the face of God. Voegelin wrote,

> What is new in the eleventh and tenth centuries of Israelite history is the application of psychological knowledge to the understanding of personalities who, as individuals, have become the carriers of a spiritual force on the scene of pragmatic history. No such character portraits [as those found on page after page in the Old Testament] were ever drawn of Babylonian, Assyrian, or Egyptian rulers, whose personalities . . . disappear behind their functions as the representatives and preservers of cosmic order in society.[5]

As Voegelin's observations suggest, the "leap in being" catalyzed by Yahweh's revelation of himself to ancient Israel opened up fundamentally new possibilities for human being-in-the-world. It would take centuries for these possibilities to sink into human consciousness, and responsible personal agency before God would not be pushed to its logical conclusion in the innermost recesses of the human soul until the revelation of Israel's Messiah. Still, ancient Israel's experience of revelation signaled the turning of a new page in human history. Thereafter "response-*able*," personal, and truly historical agency would move to center stage in world history.

The divine address that occasioned human response-*ability*, furthermore, was neither arbitrary nor capricious, but ethical and rational, established by covenant, and—most importantly for our present purposes—*fixed in written texts*. Indeed, it appears that an ingenious technology played a crucial role in facilitating Israel's "leap in being." That technology, of course, was literacy, that simple yet clever technique of objectifying thought in a durable and transmissible form. Writing and reading pull us temporarily out of the immediacy of being. They foster reflection, introspection, individuation, and the self-conscious appreciation of oneself as over and against others. By separating the knower from the known, as cultural historian Walter Ong observes in a seminal study of the historical and cultural impact of literacy, writing makes an

---

[5]Voegelin, *Israel and Revelation*, 223.

increasingly articulate introspectivity possible.[6] It opens the self to both the external objective world and its own interior world, over and against which the objective world—including other persons—is now apprehended. "More than any other single invention," Ong declares, "writing has transformed consciousness."[7] And, indeed, as he notes elsewhere, "all major advances in consciousness depend upon technological transformations and implementations of the word."[8]

The technological transformation and implementation of the word that has resulted in some of the most significant advances in human consciousness was *phonetic* literacy. Because it is relatively easy to learn and employ, phonetic literacy was surely one of the most significant technological developments in human history. It appears to have made possible a new understanding of the self, and with this new understanding, any number of new social and cultural formations.

As far as historians have been able to discover, the phonetic alphabet was invented just once—or possibly twice in rapid succession in very nearly the same location—among Northern Semitic peoples around 1500 BC.[9] The phonetic system appears to have been invented to keep track of simple commercial transactions—which is to say, for mundane practical and economic reasons—but it also came to be used to record and mediate powerful religious experiences. In conjunction with the use of the alphabet, a new space seems to have opened up between human culture and the immediacy of nature. This was perhaps because, as philosopher David Abram notes, the written characters of the phonetic alphabet—its phonemes—no longer correspond to sensible phenomena but solely to the range of sounds the human mouth can produce.[10] In using the alphabet, our attention is subtly shifted away, Abram reasoned, "from any outward or worldly reference of the pictorial image . . . [and a] direct association is established between the pictorial sign and the vocal gesture, for the first time completely bypassing the thing pictured."[11]

---

[6]Walter J. Ong, *Orality and Literacy: The Technologizing of the Word* (London: Methuen, 1982), 105.
[7]Ibid., 78.
[8]Walter J. Ong, *Interfaces of the Word: Studies in the Evolution of Consciousness and Culture* (Ithaca, NY: Cornell University Press, 1977), 42.
[9]Walter J. Ong, *The Presence of the Word* (New Haven, CT: Yale University Press, 1967), 39.
[10]David Abram, *The Spell of the Sensuous* (New York: Vintage, 1996), 100.
[11]Ibid., 100-101.

Abram laments this development, believing that the roots of our modern ecological crisis may be traced back to this separation of the human from the natural. He suggests, furthermore, that the silencing of nature has become particularly problematic within Christian civilization precisely because Christianity's original documents were written in Greek, the first language in which both consonants and vowels were represented by abstract phonetic symbols. Abram's provocative thesis is reminiscent of Marshall McLuhan's contention in *Understanding Media* that phonetic literacy lies at the root of Western technical rationality. "Only alphabetic cultures," McLuhan wrote,

> have ever mastered connected lineal sequences as pervasive forms of psychic and social organization. The breaking up of experience into uniform units in order to produce faster action and change of form (applied knowledge) has been the secret of Western power over man and nature alike. That is the reason why our Western industrial programs have quite involuntarily been so militant, and our military programs have been so industrial. Both are shaped by the alphabet in their technique of transformation and control by making all situations uniform and continuous.[12]

Now, although Abram's indictment of Christian civilization is debatable, and while McLuhan seems often to have allowed himself to be carried away by his own rhetoric, phonetic literacy does appear to have occasioned a number of fundamentally new human possibilities that the inventors of this "transformation and implementation of the word" could not have foreseen. This ingenious technological invention, in short, seems closely bound up with what has traditionally been assumed to be an exclusively theological development, an understanding that places a great deal of emphasis upon responsible and personal human agency before God and neighbor—which is to say, of *love*. While no one confesses "Jesus is Lord" except by the Holy Spirit (1 Cor 12:3), nevertheless beyond that first generation of eyewitnesses, Christians everywhere have come to faith on the basis of the apostles' testimony to Jesus. It pleased God that their testimony should have been preserved with the aid of phonetic literacy.

---

[12]Marshall McLuhan, *Understanding Media: The Extensions of Man* (New York: Signet, 1964), 88.

Automatic Machine Technology

And so it is surely no exaggeration to say that technological making is an essential aspect of human being. As the above discussion of phonetic literacy's cultural, social, psychological, and even spiritual significance suggests, our present understanding of ourselves and our estimation of our potential as human beings have been decisively shaped by technologies, many of which we take almost entirely for granted, and whose inventors and histories have been largely forgotten. To question the continued progress of modern technology would therefore seem to be at odds with human historical development. It would seem to foreclose upon the possibility of human consciousness's further technology-aided evolution. After all, had we heeded Socrates's admonitions about the debilitating impact that literacy would have upon memory,[13] we might never have experienced Voegelin's "leap in being." We might never have developed the complex interiority that was so necessary to the thoughtful and responsible personal agency we have come to prize.

Still, prudence requires us to at least ask whether our technologies are now enabling us to become more of ourselves; whether they are facilitating deeper and more meaningful human relationships; whether they are enabling us to be more dynamically engaged with the world. One suspects that they are not. Indeed, it appears that some of our technologies are actually—albeit inadvertently—*dis*-engaging us from each other, from the natural world, from ourselves—and indeed, as Albert Borgmann puts it, from skilled and bodily commerce with reality itself.[14]

As we mentioned in the introduction, Borgmann attributes this problem to our increasing reliance upon technological mechanisms, systems, and gadgets that promise to "disburden" us, making our lives easier, more convenient, safer, more pleasurable, and so on. Borgmann considers the rise of what he calls "the device paradigm" to be the most consequential development of the modern period, explaining everything from social injustice

---

[13]Plato, "Phaedrus," in *The Collected Dialogues of Plato Including the Letters*, ed. Edith Hamilton and Huntington Cairns, Bollingen Series 71 (Princeton, NJ: Princeton University Press, 1964), 520.

[14]Albert Borgmann, *Technology and the Character of Contemporary Life: A Philosophical Inquiry* (Chicago: University of Chicago Press, 1984), 151.

to political indecision and the destruction of the natural environment.[15] We will return to Borgmann's analysis, but it will suffice here to say that one of the crucial features of modern technological devices is not simply that they demand little from us beyond our time and money. They also produce and deliver their commodities to us more or less automatically, often by way of quite complicated—if hidden—machinery. Modern devices, in short, are examples of automatic machine technology. They have been expressly designed and developed to do much of what they do mechanically, automatically, and without need of skilled human input.

While it has often been the case that our technologies have empowered us to become more of ourselves, modern automatic machine technology seems rather to allow us to become less, diminishing us even as it purports to deliver "more" and "better," "faster" and "easier," "new" and "improved." Again, this is not terribly surprising. For beyond supplying us with power and commodities, automatic machine technology has been designed, developed, and deployed precisely to eliminate much, if not all, that is characteristically human from the technological process. This may seem to be an extreme statement, but it is actually rather obvious once we understand what machines are.

One of the key characteristics of machine technology, Lewis Mumford pointed out in his classic work on the history of technology, *Technics and Civilization*, is automatism.[16] Machines are designed and developed to function automatically and self-sufficiently, unimpeded—as much as possible—by human frailties, inconsistencies, and irrationalities. Indeed, automatism is one of the keys to their utility. Karl Marx was among the first to comment on this in the early decades of the nineteenth century. The world-historical importance of distinctively modern industry, Marx observed, was that it had succeeded in breaking the productive process down into a series of discrete operations and separately analyzable steps, many of which could be automated. The modern system thereby freed itself from the capricious, unpredictable, and unreliable influence of human judgment. While human beings continued to be involved in the industrial process—as

---

[15]Ibid., 3.
[16]Lewis Mumford, *Technics and Civilization* (London: Routledge & Sons, 1934), 10.

owners, engineers, managers, or workers—the influence of individual human beings *as such* had been minimized within what would come to be called "the factory system." Only after individual human influence had been thus minimized was industry able to take full advantage of modern scientific knowledge. Having reached this stage, Marx contended rather optimistically, modern machine technology had become capable of indefinite improvement. Indeed, all that remained to be accomplished was for the relations of production—between the factory system's owners and workers—to catch up to the system's remarkable productivity.[17]

The emergence of modern scientific knowledge is the other key to understanding modern machine technology. For it is in modern machinery's incorporation of the insights of modern science that it differs from tools per se, as well as from earlier human technologies. In an essay entitled "The Machine and Humanity," Romano Guardini provided a helpful framework for understanding just how "machines" differ from "tools" and even from more elaborate "contrivances."[18] The key lies in the relation of each to "given nature." Guardini prefaced his argument by contending—in a way similar to Voegelin's observations above—that genuinely human culture entails a measure of distance from the immediacy of given nature. It is more authentically human, for example, to live in a house than to live in a cave. Yet Guardini insisted that authentic human culture needed to remain in living conversation with nature *as given*. A human dwelling that is carefully fit into the natural landscape, from his perspective, is more authentically human than a unit in a subdivision that has been forcibly imposed upon a landscape by means of earth-moving machinery.

A "tool," Guardini suggested, is something that enhances what the members and organs of our bodies can more-or-less naturally accomplish.[19] A hammer, for example, is a tool that amplifies and focuses the force that we might otherwise have to apply with a rock or, failing that, with our fist. A "contrivance," continuing on—such as a waterwheel or windmill—is an

---

[17]See Nathan Rosenberg's discussion of Marx in *Inside the Black Box: Technology and Economics* (Cambridge: Cambridge University Press, 1982), 34.

[18]Romano Guardini, "The Machine and Humanity," in *Letters from Lake Como: Explorations in Technology and the Human Race* (Grand Rapids: Eerdmans, 1994 [1923]), 97.

[19]Ibid., 98.

assemblage of tools that extends well beyond ordinary bodily functions, yet nevertheless operates because natural forces work directly upon it.[20] For example, when there is not enough water to turn its wheel, a waterwheel ceases to function. It must wait upon given nature for its continuing operation. Tools and contrivances thus rest on natural processes that can be immediately experienced. Yes, they both make use of nature, but they make use of nature as it offers itself up for our use. "This field of operation of tools and contrivances," Guardini wrote, "leads without break to directly existing nature on the one hand and to directly given humanity on the other."[21]

A "machine," by contrast, is different. It displays a fundamentally new relation to nature, a relation only made possible in the modern period by careful scientific analysis. Modern science obviously seeks to know and understand nature, but it is not typically concerned with how given nature presents itself to us in the ordinary course of events. Rather, science seeks to discover nature's inner workings, analyze its constituent parts, and determine the forces and laws by which it operates and develops. Modern machine technology, then, takes advantage of precisely this kind of scientific knowledge. A "machine" may be said to be present, Guardini noted, "only when the function is scientifically understood and technically worked out so that the mode of operation can be accurately fixed."[22] As William Rosen details in a lively history of the machine that powered the Industrial Revolution, *The Most Powerful Idea in the World*, the development of the modern steam engine depended upon the practical application of dozens of scientific propositions.[23] Indeed, as Rosen's study clearly indicates, the "Machine Age" only really began as the wall between scientific theory and practice that had stood for centuries was finally broken down.[24] We will return to this point in our third chapter. Suffice it to say here that it is the practical application of scientific knowledge that makes machine technology so powerful. It is also what makes machine technology so potentially destructive of given nature. Guardini observes:

---

[20]Ibid., 99.
[21]Ibid., 101.
[22]Ibid., 100.
[23]William Rosen, *The Most Powerful Idea in the World: A Story of Steam, Industry, and Invention* (New York: Random House, 2010).
[24]Ibid, 74.

Modern physics and chemistry have established this kind of mastery over materials and forces. The latter must obey. What holds sway is not a vital and sympathetic power, an ability to follow the inner courses of reality and to shape it accordingly. Instead, mechanistic laws are present that have been established once and for all and that anyone can manipulate. Materials and forces are harnessed, unleashed, burst open, altered, and directed at will. There is no feeling for what is organically possible or tolerable in any living sense. No sense of natural proportions determines the approach. A rationally constructed and arbitrarily set goal reigns supreme. On the basis of a known formula, materials and forces are put into the required condition: machines.[25]

A machine, Guardini concludes, is a kind of "iron formula" designed to direct materials and forces to humanly desired and determined ends.[26]

As Guardini's comments indicate, premodern human culture could be described as a kind of dialogue between human ingenuity and given nature. Modern machine technology by contrast represents something of a monologue. The only active participants in the modern conversation are those who own, design, and/or use the machinery, and they are not so much concerned to listen and respond to given nature as they are to put the forces and materials of nature, revealed by modern science, to good use. Modern machine technology thus represents the synthesis of technical ingenuity with scientific knowing, a synthesis in which both activities are mightily amplified. As Grant once put it, "Modern technology is not simply an extension of human making through the power of a perfected science . . . [it] is a new account of what it is to know and to make in which both activities are changed by their co-penetration."[27] This new account lies at the very heart of modern industrial civilization.

Of course, this does not mean we cannot and should not be deeply grateful for modern machine technology. For, as Guardini went on to observe, machines have enabled us to achieve "ever higher and more nuanced goals with ever greater certainty."[28] And yet as Guardini, Ellul, Grant, and a

---

[25]Guardini, *Letters from Lake Como*, 46.

[26]Ibid.

[27]George Grant, "Thinking About Technology," in *Technology and Justice* (Toronto: Anansi, 1986), 13.

[28]Guardini, "The Machine and Humanity," 102.

great many others have lamented, the proliferation of modern machine technology has also resulted in the loss of a vital connection with nature, as well as in the loss of humane wisdom. In a vivid example, Guardini urges his readers to contrast a traditional sailing vessel, in itself a remarkable and ancient cultural and technological achievement, with a modern ocean liner.[29] Both clearly testify to the human mastery of nature, yet the former necessarily remains in close "conversation" with the natural forces of wind and water. The latter, by contrast, has become so large that, beyond the physics of buoyancy and hull speed, given nature no longer has any real power over it and can no longer be seen on it. Those aboard, Guardini laments, "live as if in houses and on city streets." "Not only has there been step-by-step development, improvement, and increase in size," Guardini writes, "a fluid line has been crossed that we cannot fix precisely but can only detect when we have long since passed over it—a line on the far side of which living closeness to nature has been lost."[30] With its unremitting emphases upon efficiency, continuous motion, predictability and prearranged effects, standardized measurements, interchangeable parts, and so forth, modern machine technology seems to have produced a world that is often artificial and occasionally even strangely alien to human feeling.

This comes as no great surprise for, as we have said, machines have from the beginning been designed, developed, and deployed to function self-sufficiently and independently of all but a minimum of human input. This, again, is one of the main reasons modern machines work so efficiently and reliably. Yet it is also why we should not expect modern machine technology, in and of itself, to open up new avenues for human growth and development. While automatic machine technology confers enormous power and control—particularly upon those who own and manage it—it has not enabled, is not now enabling, and will not *ever* enable the rest of us to become more of ourselves except insofar as it supports the material basis of our lives. Should it, for example, occur to me to think about the machine systems that enable me to procure fresh strawberries in the dead of winter and at practically any time of the day or night, I should surely be amazed

---

[29]Ibid., 11.
[30]Ibid., 13.

and perhaps even grateful, but I don't in any meaningful sense become more of myself in purchasing strawberries at my local Safeway. Rather, I simply play my small part within the larger system as a consumer. Hence, although the power and control conferred by modern machine technology must not be taken for granted, we should nevertheless not be deceived by it into turning a blind eye to the threat that automatic machine technology now poses to us and to our world.

In this connection, we have unfortunately tended to conceive of our relation to modern machine technology on the analogy of a master's relation to a slave. Presumably, our machines don't mind being enslaved. This has tended to give rise to a fundamentally optimistic view of technological progress. As Carr recently commented,

> If we assume that our tools act as slaves on our behalf, always working in our best interest, then any attempt to place limits on technology becomes hard to defend. Each advance grants us greater freedom and takes a stride closer to, if not utopia, then at least the best of all possible worlds. Any misstep, we tell ourselves, will be quickly corrected by subsequent innovations.[31]

Unfortunately, this optimistic and typically modern view overlooks the fact that the master/slave relation is nearly as dehumanizing for the master as it is for the slave. We gain precious little self-knowledge in our relations to and use of our mechanical slaves, and ominously, the relation often seems to become inverted. We often end up, in effect, becoming slaves to mechanical masters. Consider our recurrent use and utter dependence upon clocks. As Mumford famously observed, the gains in productivity made possible by the clock's "closer articulation of the day's events" can hardly be overestimated.[32] Yet modern men and women typically complain that they are relentlessly driven by schedules, appointments, timetables, and agendas. In the last century, Western visitors to the Philippines were described as "people with gods on their wrists."[33] Today, of course, we hold our gods in the palms of our hands. "On average," one recent study found, "people in the

---

[31]Nicholas Carr, *The Glass Cage: Automation and Us* (New York: W. W. Norton, 2014), 226.

[32]Mumford, *Technics and Civilization*, 17.

[33]Cited in Os Guinness, *The Last Christian on Earth: Uncover the Enemy's Plot to Undermine the Church* (Ventura, CA: Regal, 2010), 50.

United States across all age groups check their phones 46 times per day,"
roughly once every fifteen minutes. For people between the ages of eighteen
and twenty four, that number goes up to seventy four times per day, or once
every twelve minutes.[34] While calling this a "slavery" may be an exaggeration, it would seem at the very least to disclose a form of dependency.

As we have built upon analytical methods and procedures, furthermore,
our use of machine technology has fostered an outlook that tends to
transform our world into an exclusively "human-made" environment, comprised almost entirely of human artifacts on the one hand and potentially
manipulable material on the other. In a celebrated short essay in which he
sought to identify the essence of modern technology, philosopher Martin
Heidegger lamented that the "impression comes to prevail that everything
man encounters exists only insofar as it is his construct."[35] This, Heidegger
continued, has given rise to a kind of delusion in which "it seems as though
man everywhere and always encounters only himself."[36] Yet human beings
can only ever really "encounter" themselves by entering into genuine conversations with "others," including the world of nature.

We will return to Heidegger's intriguing analysis in our third chapter.
Here, we simply affirm that modern machine technology's power over
nature is the source of its great appeal but also of its insidious potential for
dehumanization. We will concern ourselves with this problem in the remainder of this first chapter. Basically, I want to consider some of the evidence suggesting that automatic machine technology's impact upon us is
not altogether beneficial; that there is, in other words, a problem, and that
the particular shape of the problem is directly related to modern automatic
machine technology.

---

[34]Lisa Eadicicco, "Americans Check Their Phones 8 Billion Times a Day," *Time*, December 15, 2015,
time.com/4147614/smartphone-usage-us-2015/. Of course, these 2015 figures are almost certainly
outdated. A recent *Globe & Mail* article (January 2018) reports that recent estimates suggest that
average smartphone users check their phones on average upward of 150 times per day; see Eric
Andrew-Gee, "Your Smarphone Is Making You Stupid, Antisocial, and Unhealthy. So Why Can't
You Put It Down?," www.theglobeandmail.com/technology/your-smartphone-is-making-you-
stupid/article37511900.

[35]Martin Heidegger, "The Question Concerning Technology," in *The Question Concerning Technology and Other Essays*, trans. William Lovitt (New York: Harper & Row, 1977), 27.

[36]Ibid.

The "Cognitive Style" Fostered by Modern Technology

In *The Homeless Mind: Modernization and Consciousness*, sociologists Peter and Brigitte Berger, along with Hansfried Kellner, sought to describe the unusual cognitive style required and encouraged within the typically modern workplace.[37] The authors found that because the modern workplace tends to be organized around machines and machine systems, and because the actions of individual workers are themselves intrinsic to the machinery, it encourages workers to think of themselves (as well as each other) as reproducible "components" within the work process. Each worker's contribution to the process can be precisely measured and thereby "objectively" stipulated, and no particular role within the process is unique and/or irreplaceable, so that any worker with comparable training can readily substitute for any other worker within the system. Workers are thus encouraged to think of the work process as something that can be broken out into steps, procedures, assemblies, components, and so on.[38] Not surprisingly, Berger et al. found that this tends to encourage somewhat anonymous social relations within the work environment. Indeed, they noted that the modern workplace tends to foster even a kind of "self-anonymization."[39]

Most of those involved in the modern work process only possess a rudimentary understanding of how all the various steps, procedures, assemblies, components, and roles fit together to make up the larger system. Still, they trust that expert knowledge of the larger system and all its individual aspects can be made available if needed.[40] Presumably there are managers who understand and control the larger process. What this reliance upon "expertise" implies, however, is that individual workers, though often highly trained in specialized skills and disciplines, are all but forced to yield to a kind of separation of means from ends. "Since reality is apprehended in terms of components which can be assembled in different ways," Berger and colleagues observed,

---

[37]Peter L. Berger, Brigitte Berger, and Hansfried Kellner, *The Homeless Mind: Modernization and Consciousness* (New York: Vintage, 1974).
[38]Ibid., 27.
[39]Ibid., 33.
[40]Ibid., 25.

there is no necessary relationship between a particular sequence of compo-
nential actions and the ultimate end of these actions. To take an obvious example, a particular assemblage of cogs produced in a highly specific production sequence may eventually go into a passenger automobile or a nuclear
weapon. Regardless of whether the worker involved in this particular production process approves or even knows about its intended end, he is able to
perform the actions that are technologically necessary to bring it about.[41]

In short, the authors found that the modern workplace tends to discourage
"big picture" thinking about the meaning and significance of the work process.

On the other hand, the modern workplace does encourage what Berger
et al. termed "problem-solving inventiveness," or what is more commonly
called a kind of "engineering mentality." The engineering mentality assumes
that any problem can, with skill and determination, be converted into a
technical problem, and that once it has thus been converted it thereby becomes amenable to a technical solution. And just as the engineering mentality was initially applied to the manipulation of nature and machines, so
we have now brought this outlook to bear on social and political life. Indeed,
the authors suggested that it is not uncommon today for individuals to seek,
often by way of the expert advice of those trained in modern psychology, to
"engineer" themselves.[42]

Not surprisingly, when combined with the modern workplace's separation of means from ends, problem-solving inventiveness and the engineering mentality often seem blind to the overall impact that machine technology has made upon our world. Again, Grant's observations cited in the
introduction are apropos: the very cognitive habits that have made us leaders
in the development of technology now impede our ability to comprehend
technique from beyond its own dynamism.[43] As the old tag goes, we cannot
see the forest for the trees.

Lastly, Berger et al. observed that the anonymity and impersonality of the
public workplace frustrate the quest for individual and personal identity.

[41]Ibid., 27-28.
[42]Peter L. Berger, *Pyramids of Sacrifice: Political Ethics and Social Change* (Garden City, NY: Anchor, 1974), 20.
[43]George Grant, "In Defense of North America," in *Technology and Empire: Perspectives on North America* (Toronto: Anansi, 1969), 40.

Under modern conditions, if one is really to "find oneself" (as the quest for identity is often described), one must not conduct the search in the public world of work so much as in what has come to be called the private sphere of life. Berger and colleagues observed that one of the more important consequences of the Industrial Revolution was a kind of institutional segregation of "public" and "private," and that a similar disjunction has occurred at the level of consciousness, with important consequences for modern society and culture.[44] Of course, modern individuals often take the problem-solving inventiveness they have acquired in the public workplace back with them into the private sphere, which predisposes them to employ techniques and technologies in their ongoing quest for identity. We will return to this point below.

Technology and the "De-skilling" of Labor

In reaction to the above observations, it often is suggested that although assembly line or factory floor work may be somewhat mind numbing, this is not the kind of work that most of us actually do today. Modern technology has freed many of us to pursue newer, more interesting, and more stimulating occupations. Indeed, new vocational opportunities requiring the development of new and challenging technical skills (so the argument runs) have more than made up for whatever skills may have been lost after, say, the last wagon wheel was hand fabricated.

Yet while this may be true for the relatively few who occupy positions at the leading edges of technological development, it does not actually appear to be true for most of the rest of us. For just as we have learned to maximize efficiency by breaking down the work process into steps and procedures that anyone with the requisite training can perform, so we have also learned to do with a minimum of skilled human labor, which, after all, is expensive and not always easily and immediately replaceable. For example, Borgmann observed, "The leading work inevitably leaves in its wake a wasteland of divided and stultifying labor. Wherever there is a traditional area of skillful work, it is disassembled, reconstructed, largely turned over to machines, and

---

[44]Berger, Berger, and Kellner, *Homeless Mind*, 29-30.

artisans are replaced by unskilled laborers. This process leads to a continuing contraction of expertise and a corresponding expansion of unskilled labor."[45]

Borgmann is by no means alone in drawing attention to the growth of unskilled labor within the modern technological milieu. Nicholas Carr has discussed the problem in several recent studies.[46] Carr begins *The Glass Cage: Automation and Us* (2014) by recounting the ironic contribution of automated systems to several recent airline disasters in which it was determined that human pilots, having become so used to computer control, had basically forgotten how to fly their aircraft.[47] A similar diminishment of proficiency appears to extend to *anyone* who uses *any* kind of automated device for *any* purpose. For example, Carr describes the de-skilling impact that GPS devices have had upon the Inuit in the far north of Canada. Tragically, the remarkable wayfinding skills that the Inuit have long been known for appear to be disappearing with the use of the new devices. Rather than augmenting their legendary ability to sense just where they are in the featureless wilderness in which they hunt, their use of GPS technology has led to the rapid loss of traditional skills. As a result, Carr writes, "A singular talent that has defined and distinguished a people for thousands of years may well evaporate over the course of a generation or two."[48] Once the Inuit's wayfinding talents have been lost, it is difficult to imagine how they could be resuscitated.

In seeking to shed light on the de-skilling of modern labor, Carr cites the rigorous examination of automation's effects upon skill levels undertaken by James Bright of the Harvard Business School in the 1950s. Bright examined the consequences of automation on workers in thirteen different industrial settings. He found that skill levels increased marginally with the introduction of power hand tools, but they invariably dropped as more complex machinery was introduced. Skill levels dropped off particularly sharply as workers began to use automated, self-regulating machinery.[49]

---

[45]Borgmann, *Technology and the Character of Contemporary Life*, 118.

[46]See Nicholas Carr, *The Shallows: What the Internet Is Doing to Our Brains* (New York: W. W. Norton, 2010); Nicholas Carr, *The Big Switch: Rewiring the World from Edison to Google* (New York: W. W. Norton, 2013); and Carr, *The Glass Cage*.

[47]Carr, *The Glass Cage*, 43.

[48]Ibid., 127.

[49]Ibid., 110-11.

Carr is perhaps best known for posing the provocative question in his 2008 *Atlantic* piece "Is Google Making Us Stupid?"[50] Insofar as we value rigorous, focused, deliberate, considered, independent, and original thinking, his answer is simple: "Yes, it almost certainly is." For, as he notes in *The Big Switch*, the Internet simply has not been designed—or at least not since it became a predominantly commercial matrix—to promote slow, deliberate, sustained, and disciplined thought. Rather, it is being designed and painstakingly developed to encourage "immediacy, simultaneity, contingency, subjectivity, disposability, and, above all, *speed*."[51] Futurologist Richard Watson observes that the kind of deep thinking inherent in strategic planning, scientific discovery, and/or artistic creation simply cannot be done in "an environment full of interruptions or hyperlinks. It can't be done in 140 characters. It can't be done when you're in multitasking mayhem."[52]

The question we currently face, as Watson correctly points out, is whether, as our machines become increasingly intelligent and human-like, we will simply lapse into ignorance and become more and more machine-like.[53] Carr, citing Vivek Haldar, captures this trade-off in the quip "Sharp tools, dull minds."[54] And the outlook is not particularly encouraging. As one of Facebook's engineers is reported to have lamented recently, "The best minds of my generation are thinking about how to make people click ads."[55]

Somewhat more disturbingly, Carr reviews recent research that has sought to explain the neurobiological mechanisms that seem to be responsible for the "degeneration effect" reflected in the loss of skill.[56] Basically, it appears that when automated systems and devices come between us and the world of lived experience, we inadvertently short circuit the neurobiology that allows our brains and bodies to be changed by experience. In other words, as automated systems relieve us of repetitive mental exercise—which, of course, is precisely what these systems have been designed to do—they

---

[50]Nicholas Carr, "Is Google Making Us Stupid?" *Atlantic*, July/August 2008.
[51]Carr, *The Big Switch*, 228.
[52]Richard Watson, *Future Minds: How the Digital Age Is Changing Our Minds, Why This Matters, and What We Can Do About It* (London: Nicholas Brealey, 2010), 4.
[53]Ibid., 50-51.
[54] Vivek Haldar, cited in Carr, *The Glass Cage*, 78.
[55]Cited in Andrew Keen, *The Internet Is Not the Answer* (New York: Atlantic Monthly Press, 2015), 60.
[56]Carr, *The Glass Cage*, 65.

undermine the possibility of learning. We realize this when confronted with the need to calculate a value without a calculator or smartphone. Simple mathematical operations that would once have been routine have apparently—and disconcertingly—been lost to disuse. Here we are reminded of the question nicely posed by historian George Dyson in 2008: "What if the cost of machines that think is people who don't?"[57] or, we might add, people who *can't*?

Putting this in terms we employed above, automated systems and devices interfere with the conversations that we apparently need to have with reality in order to grow and develop. As clever, as effective, and as useful as our devices may be, they can nevertheless diminish us, rendering our lives somewhat less than real. When we allow automation to get between us and the world, Carr warns, it tends to erase the artistry from our lives.[58] Along this line, the American composer and conductor John Philip Sousa predicted in 1906 that the technology of musical recording would inevitably lead to the demise of music. "The time is coming," Sousa wrote, "when no one will be ready to submit himself to the ennobling discipline of learning music. Everyone will have their ready made or ready pirated music in their cupboards."[59] While Sousa was clearly mistaken about music per se, recorded music does tend to obscure the human effort that has gone into making it. As Alex Ross commented some years ago in *The New Yorker*, the paradox of musical recording is that while it is able to capture moments of musical sound, it fails to preserve the human spirit with which the music was produced.[60]

"Technological Unemployment"

As machinelike as we have managed to become in our efforts to accommodate the machinery that pervades the modern workplace, we cannot ultimately compete with the machines. They are stronger, faster, more precise, more reliable, tireless, and significantly less expensive than we are. For just these reasons, business and industry have been progressively replacing

---

[57]Ibid., 113.
[58]Ibid., 85.
[59]Sousa, cited in Ross, "The Record Effect," www.newyorker.com/magazine/2005/06/06/the-record-effect.
[60]Ibid.

human labor with capital machinery over the course of the last century and a half. The celebrated British economist John Maynard Keynes coined the term "technological unemployment" to describe this process nearly a hundred years ago. "This means," Keynes wrote in an essay entitled "Economic Possibilities for Our Grandchildren," that "unemployment due to our discovery of means of economizing the use of labour [is] outrunning the pace at which we can find new uses for labour."[61] Of course, Keynes was optimistic that new uses for human beings would eventually be found and that, in any event, the wealth generated by machine-fueled productivity would benefit everyone within the larger society—and for the most part, those two assumptions have proven true.

The cost savings typically made possible by new machine systems have led to lower prices, and lower prices have, in turn, stimulated demand that has led to new investment in production facilities, creating new opportunities for employment, and so forth. That the new machine systems also led to the production of entirely new products has amplified the compensatory effect of new technologies. And, of course, it has further been assumed that the kind of work that has been surrendered to machinery is not the kind of work that most people would choose to do anyway. At least this was the assumption until fairly recently.

Economists have lately noted that rapidly advancing information technologies now appear poised to extend the phenomenon of technological unemployment beyond the domain of the blue-collar job and into the realm of white-collar occupations.[62] Indeed, the machinery that once disrupted traditional manual production appears ready to disrupt elite professional vocations such as law, medicine, finance, and education. In a somewhat alarming study entitled *Who Owns the Future?*, philosopher and computer scientist Jeron Lanier predicts that virtually every sector of the postmodern economy will soon be digitized and made amenable to "software mediation." "Software," Lanier writes,

---

[61]John Maynard Keynes, "Economic Possibilities for Our Grandchildren," in *Essays in Persuasion*, 358-73 (New York: W. W. Norton, 1963), 3 (available at www.econ.yale.edu/smith/econ116a/keynes1.pdf).
[62]Carr, *Big Switch*, 136.

could be the final industrial revolution. It might subsume all the revolutions to come. This could start to happen, for instance, once cars and trucks are driven by software instead of human drivers, 3D printers magically turn out what had once been manufactured goods, automated heavy equipment finds and mines natural resources, and robot nurses handle the material aspects of caring for the elderly.[63]

Lanier admits that digital technology may not advance rapidly enough in the twenty-first century to dominate the economy, but his best guess is that it will, and he fears that the result will be hyper-unemployment. For when much of what human beings used to produce and do is produced and done by machines, a great many of us are going to find ourselves out of work. "As the information economy arises," Lanier predicts, "the old specter of a thousand science fiction tales and Marxist nightmares will be brought back from the dead and empowered to apocalyptic proportions. Ordinary people will be unvalued by the new economy, while those closest to the top computers will become hypervaluable."[64]

Lanier's predictions of the apocalypse are perhaps (hopefully!) far-fetched, but we are even now seeing signs of the kind of income disparity that he predicted. Economic growth and employment are diverging in advanced industrial countries, and this does indeed appear to be due to the productivity gains that have been made possible by increasingly sophisticated computer-aided machinery. As economists Erik Brynjolfsson and Andrew McAfee report, sustained exponential improvement in the speed and capacity of computers, the ongoing digitization of an extraordinarily large amount of data, and the use of increasingly sophisticated algorithms to mine this data for useful patterns are yielding technological breakthroughs that will almost certainly make the stuff of yesterday's science fiction actually possible within the coming years.[65] However, Brynjolfsson and McAfee note that while it was once assumed that increasing automation would free people

---

[63]Jaron Lanier, *Who Owns the Future?* (New York: Simon & Schuster, 2013), 7.

[64]Ibid., 15.

[65]Erik Brynjolfsson and Andrew McAfee, *The Second Machine Age: Work, Progress, and Prosperity in a Time of Brilliant Technologies* (New York: W. W. Norton, 2014), 90; see also Stanley Aronowitz and William DiFazio, *The Jobless Future*, 2nd ed. (Minneapolis: University of Minnesota Press, 2010); also Jeremy Rifkin, *The End of Work: The Decline of the Global Labor Force and the Dawn of the Post-Market Era* (New York: Penguin, 1995).

up to pursue science, philosophy, art, or travel, it is beginning to look like these privileges may only actually accrue to the few who own and/or manage the machinery, much as Lanier suggested.

The irony is that central "Second Machine Age" entities like Google and Facebook are becoming so hypervaluable precisely because of all of the information we are providing them about ourselves—*for free*. As Andrew Keen comments in *The Internet Is Not the Answer*, all of us who use Google's search services or post on Facebook are effectively working for these firms without being compensated. Their commercial viability hinges on capturing, packaging, and selling the data generated by our online lives, our so-called data exhaust.[66] This information is sold to advertisers, media outlets, and others seeking to understand, anticipate, and possibly manipulate our behavior. As Carr explains, "We are the Web's neurons, and the more links we click, pages we view, and transactions we make—the faster we fire—the more intelligence the Web collects, the more economic value it gains, and the more profit it throws off."[67] Keen cites a leading computer security expert, Bruce Schneider, as having commented that the principal business model currently at work on the Internet is based very largely on "mass surveillance."[68] We tend to assume that the Internet is something that we use, but in fact we are being used by it, or at least by those who own and control it.[69]

Returning to the issue of technological unemployment, there are a number of reasons why the "Second Machine Age," as Brynjolfsson and McAfee call it, looks to pose significant challenges to human employment. Some of those challenges have to do with the efficiencies made possible by information and communications technologies. Others have to do with consumer behavior in response to these technologies. First and most obviously,

---

[66]Keen, *The Internet Is Not the Answer*, 60.

[67]Carr, *The Big Switch*, 228.

[68]Keen, *The Internet Is Not the Answer*, 182.

[69]This, we note, is the thesis of Lanier's book, *Who Owns the Future?*, and he proposes devising systems to compensate each of us for the small contributions we make to the Internet's trove of data. "An amazing number of people offer an amazing amount of value over networks," Lanier writes (p. 9). "But the lion's share of wealth now flows to those who aggregate and route those offerings, rather than those who provide the 'raw materials.' A new kind of middle class, a more genuine, growing information economy, could come about if we could break out of the 'free information' idea and into a universal micropayment system. We might even be able to strengthen individual liberty and self-determination even when the machines get very good."

computers have gotten very good at an increasing number of things. It has long been assumed that the dexterity and reach of the human intellect would always exceed that of artificial intelligence, but recent advances in computing technology have substantially narrowed the gap. Computers, it turns out, don't have to replicate the processes of human thought so long as they can replicate—thanks largely to their capacity for sifting through immense amounts of data very, very rapidly—the *outcomes* of human thought.

For instance, a computer system named ALPHAGO, developed by Google subsidiary DeepMind, was recently able to defeat the world's best human players of Go, an ancient Chinese game so complicated that it was thought only human beings could play it. Yet by accessing vast libraries of Go matches amassed over the game's long history, as well as by playing millions of games against itself, the ALPHAGO system taught itself to play and eventually to master the game.[70] Less spectacularly but of greater consequence to human unemployment, artificially intelligent machines are increasingly able to undertake complicated legal research, issue sophisticated medical diagnoses, execute multifaceted financial transactions, and a great many other operations that until recently were assumed to require human intelligence and judgment.

Computer systems and networks have also enabled a few large firms such as Walmart, Microsoft, Amazon, and Google to dominate what has become an increasingly global marketplace. The growth of these firms is partly the result of successful branding and reflects the fact that in any market, only a small number of sellers can hope to capture buyers' limited attention. Yet the spectacular growth of these firms also suggests that capacity constraints have become increasingly irrelevant in the digital marketplace. A single producer/distributor with a website can now fill the demands of millions, perhaps even billions of customers.[71] These customers naturally prefer to purchase products deemed the best, so a single producer/distributor that establishes a reputation for offering the best products at the lowest prices becomes extremely difficult to compete with. Why would anyone shop anywhere else? The larger such firms

---

[70]Andrew McAfee and Erik Brynjolfsson, "Where Computers Defeat Humans, and Where They Can't," *The New York Times*, March 16, 2016, www.nytimes.com/2016/03/16/opinion/where-computers -defeat-humans-and-where-they-cant.html.

[71]Brynjolfsson and McAfee, *Second Machine Age*, 154.

become, the more cheaply they are able to offer their products, the more profits they are able to generate, and the more they are able to invest in consolidating their brand. The emergence of giant online firms like Amazon.com—the so-called everything store—signals what has been called a winner-take-all phenomenon in the new digital economy.[72] As Brynjolfsson and McAfee write,

> Lowering prices, the traditional refuge for second-tier products, is of little benefit for anyone whose quality is not already at or near the world's best. Digital goods have enormous economies of scale, giving the market leader a huge cost advantage and room to beat the price of any competitor while still making a good profit. Once their fixed costs are covered, each marginal unit produced costs very little to deliver.[73]

As one observer recently commented, competition in the digital economy is similar to that of the old industrial economy, but "on steroids."[74]

The relevance of all of this to employment is that the automation of production and distribution systems, combined with the elimination of second-tier producers and distributors, has led to the rise of monolithic entities that can service *a lot* of customers—and generate *a lot* of revenue—with *very few* employees. A frequently cited 2013 study from the US Institute of Local Self-Reliance (ILSR) reported that "while brick-and-mortar retailers employ 47 people for every $10 million in sales, Amazon employs just 14 people to generate the same $10 million sales revenue."[75] As an even more extreme example, when Instagram was sold to Facebook in 2012 for a billion dollars, it employed only thirteen people.

Sophisticated computer systems and digital networks have also given rise to the on-demand or gig economy, in which individuals offering services are quickly, conveniently, and inexpensively linked to those desiring to purchase said services. The taxicab substitute Uber is probably the best known of this new type of on-demand firm because of the opposition it has attracted from traditional taxicab drivers and labor unions. Yet other on-demand service providers are springing up with increasing frequency. In addition to eliminating

---

[72]See Robert H. Frank and Philip J. Cook, *The Winner-Take-All Society* (New York: Free Press, 1995).
[73]Brynjolfsson and McAfee, *Second Machine Age*, 155.
[74]Keen, *The Internet Is Not the Answer*, 47.
[75]Ibid., 49.

the need for dispatchers and/or managers, on-demand systems cut human labor costs because the people offering their services through these systems are not employees in any traditional sense. They have no job security, and they are not, as yet, offered much in the way of benefits. The result, as one writer for *The Economist* recently noted, is that the on-demand economy tends to employ those who value flexibility over security, such as students, young mothers, the semi-retired, and others desiring simply to supplement their incomes. On-demand employment does not work out nearly as well for people with children to educate and/or mortgages to pay.[76]

Computers and networks have also inadvertently revolutionized any number of sectors of the postindustrial economy via online piracy. Basically, any content that can be digitized—which has thus far only included text, music, and video—can be quickly, cheaply, and illegally copied and distributed. The impact of piracy upon employment in a number of industries has been vast. As Keen notes,

> According to a 2011 report by the London-based International Federation of the Phonographic Industry (IFPI), an estimated 1.2 million European jobs would be destroyed by 2015 in the Continent's recorded music, movie, publishing, and photography industries because of online piracy, adding up to $240 billion in lost revenues between 2008 and 2015.[77]

Between 2002 and 2012, Keen goes on to note, the US Bureau of Labor Statistics reported a 45 percent drop in the number of professional working musicians, falling from over fifty thousand to around thirty thousand.[78] Much of this drop, it seems, was due to the impact of online piracy. Given that the cost of 3-D printing technology has now dropped within the reach

---

[76]"There's an App for That," *The Economist*, January 3, 2015, 17-20; a recent study reports that the "on-demand" or "gig" economy has contributed to what has been "a significant rise in the incidence of alternative work arrangements in the U.S. economy from 2005 to 2015." Indeed virtually all of the net employment growth in the US during that period appears to have been due to "alternative work arrangements." While online intermediaries like Uber only account for a small fraction of this growth, technology does appear to have reduced the transaction costs associated with contracting out job tasks, thus enabling firms to reduce the number of core employees. See Lawrence F. Katz and Alan B. Krueger, "The Rise and Nature of Alternative Work Arrangements in the United States, 1995–2015," Princeton.edu, March 29, 2016, https://krueger.princeton.edu/sites/default/files/akrueger/files/katz_krueger_cws_-_march_29_20165.pdf.
[77]Keen, *Not the Answer*, 130.
[78]Ibid., 131.

of most ordinary consumers, a great many other things besides text, music, and video may soon be available for download, both legally and illegally. This does not bode well for employment.

In sum: just as steam and electricity, as general-purpose technologies, revolutionized any number of industries, so information and communications technologies either already have disrupted—to use the currently fashionable term—or are poised now to disrupt nearly every aspect of the postindustrial economy. Brynjolfsson and McAfee's Second Machine Age will no doubt be brilliant in any number of respects, but it is also going to leave some, and perhaps quite lot of, people behind,[79] particularly the unskilled and less educated, especially overseas.[80] The authors therefore urge policy makers to develop strategies to encourage people to learn how to work with the machines because people are simply not going to be able to compete against them.[81] Other economists have suggested the implementation of a "universal base income" (UBI) or basic income guarantee that would help to compensate workers displaced by new technologies.[82] In this connection, we note that the US government has been concerned about the problem of technological unemployment since the Great Depression of the 1930s but has yet to devise effective strategies to redress it.[83]

Given that machines are expressly designed to function automatically and independently, and given our remarkable flair for designing and developing

---

[79] A 2013 study suggests that as much as 47 percent of total US employment could be at risk within the next twenty years. See Carl Benedikt Frey and Michael A. Osborne, "The Future of Employment: How Susceptible Are Jobs to Computerisation?," Oxford Martin Programme on the Impacts of Future Technology "Machines and Employment" Workshop, September 17, 2013, www.oxfordmartin.ox.ac.uk/downloads/academic/The_Future_of_Employment.pdf.

[80] Brynjolfsson and McAfee, *Second Machine Age*, 184. Here we note that recent economic data on such things as productivity, unemployment, job churn, etc. do not (yet?) reflect significant technological unemployment. "If I had to do it over again," McAfee recently commented with reference to *The Second Machine Age*, "I would put more emphasis on the way technology leads to structural changes in the economy and less on jobs, jobs, jobs. The central phenomenon is not net job loss. It's the shift in the kinds of jobs that are available" (McAfee, cited in James Surowiecki, "Roboapocalypse Not," *Wired*, September 2017, 64).

[81] This is the thesis of Brynjolfsson and McAfee's earlier book *Race Against the Machine: How the Digital Revolution Is Accelerating Innovation, Driving Productivity, and Irreversibly Transforming Employment and the Economy* (Lexington, MA: Digital Frontier Press, 2011).

[82] See, for example, Philippe Van Parijs and Yannick Vanderborght, *Basic Income: A Radical Proposal for a Free Society and Sane Economy* (Cambridge, MA: Harvard University Press, 2017).

[83] See Sue Halpern's review of Nicholas Carr's *The Glass Cage*: "How the Robots and Algorithms Are Taking Over," *The New York Review of Books* 62, no. 6, April 2, 2015, 24-28.

them, it was perhaps inevitable that we would be overtaken by the phenomenon of technological unemployment. Hitherto, this has simply meant finding new kinds of work for human beings to do. This is going to become increasingly difficult, however, for our machinery is getting to be increasingly good. The human spirit may well be willing to participate in and to take advantage of the newly automated economy, but the human flesh may simply be too weak to do so.

Technology and Private Life

In a perceptive book titled *Man in the Age of Technology*, social philosopher Arnold Gehlen observed that within the rationalized and bureaucratized modern "social system" there is little doubt but that "society expects the person to develop, to a large extent, into a 'functionary.'"[84] Those things that are distinctive about a given person, Gehlen maintained, are largely unappreciated, as are personal characteristics that hinder effective functioning.[85] Recalling Berger and colleagues' analysis above, this has tended to mean that modern individuals are forced to fashion their own personal identities largely after hours, at home and in private. The modern social situation is such that, as Berger and colleagues put it, "there *must* be a private world in which the individual can express those elements of subjective identity which must be denied in the work situation."[86]

A private sphere has indeed arisen within modern technological society, comprising primarily the family and private associations. The private sphere is the province of leisure and avocation, in which individuals are encouraged to realize all of those human things that the increasingly impersonal public workplace prohibits them from exploring and developing. In contrast with the conformity so often demanded in the workplace, in private, individuals are left almost entirely free to fabricate their own distinctive identities. The private sphere, as Berger et al. note, is "a kind of 'do-it-yourself' universe."[87] It is experienced, as Peter

---

[84]Arnold Gehlen, *Man in the Age of Technology*, trans. Patricia Lipscomb, European Perspectives (New York: Columbia University Press, 1980), 147.
[85]Ibid.
[86]Berger, Berger, and Kellner, *Homeless Mind*, 35 (emphasis in original).
[87]Ibid., 186.

Berger puts it elsewhere, as the "single most important area for the discovery and actualization of meaning and identity."[88]

The problem, of course, is that machine technology has colonized the private world too. Keen, for example, reports that in every minute of every day in 2014, "3 billion Internet users in the world sent 204 million emails, uploaded 72 hours of new YouTube videos, made over 4 million Google searches, shared 2,460,000 pieces of Facebook content, downloaded 48,000 Apple apps, spent $83,000 on Amazon, tweeted 277,000 messages, and posted 216,000 new Instagram photos."[89] It is predicted that by 2020 we will be surrounded by fifty billion networked and "intelligent" devices, in our homes and cars, in our clothing, on our roads, in the products we consume, and so on.[90] The days of making a private escape beyond the reach of technology appear to be numbered. Even in the wilderness one encounters hikers accessing Facebook and Instagram via cellular networks, and it is now considered irresponsible to enter the backcountry without a cellphone and a GPS mapping device. It should come as no surprise to learn, then, that many recent studies examining the potentially harmful effects of modern technology focus on the private use of technological devices.

In the first instance, our devices and applications, even those expressly intended to enhance sociability, appear to be leaving us lonelier and feeling more and more disconnected from one another. This is perhaps simply because they interfere with face-to-face human contact. Stories are legion of people who are apparently walking or sitting together but who are actually alone, entirely absorbed by whatever is appearing on the screens of their so-called smartphones. Why people appear to prefer to filter their relationships through such devices is something of a mystery, though it may to have to do with their desire to remain in control of their presentation of themselves, which is much more difficult to pull off in face-to-face situations. As social psychologist Sherry Turkle observes, in an interesting—if disturbing—study

[88]Peter L. Berger, *Facing Up to Modernity: Excursions in Society, Politics, and Religion* (New York: Basic Books, 1977), 133.

[89]Keen, *The Internet Is Not the Answer*, 13-14.

[90]Michael Chui, Markus Loffler, and Roger Roberts, "The Internet of Things," *McKinsey Quarterly*, March 2010 (www.mckinsey.com/industries/high-tech/our-insights/the-internet-of-things), cited by Keen, *The Internet Is Not the Answer*, 13.

of the impact of networked devices upon the young, "It is not unusual for people to feel more comfortable in an unreal place than a real one because they feel that in simulation they show their better and perhaps truer self."[91] Our "machine dream" today, as Turkle puts it, is on the one hand *never* to be alone but nevertheless *always* to remain in control of our self-projections. Such a dream simply cannot be realized in face-to-face contact with other people, so we employ our devices and the various portals of digital life to mediate between us and the others with whom we would relate.[92] "Networked," Turkle concludes, "we are together, but so lessened are our expectations of each other that we can feel utterly alone. And there is the risk that we come to see others as objects to be accessed—and only for the parts we find useful, comforting, or amusing."[93] The result, as the title of her recent book suggests, is that we can end up, in effect, being "alone together."

Given that our networked devices interfere with face-to-face relationships, and given that the Internet is painstakingly designed to distract us, it is perhaps no wonder that we have become increasingly subject to various psychological disorders of attachment and/or attention.[94] Author Michael Harris suggests that the mental state produced by our use of technological devices is one of "continuous partial attention."[95] On the one hand, he observes (citing the work of Gary Small of UCLA) that we find the voicemail, emails, instant messages, and alerts that continuously issue from our devices both stimulating and satisfying—because they indicate that we are connected to others, and even more importantly, that we are needed by them. Yet constant electronic stimulation takes a neurological toll. "In the short run," Small writes,

> stress hormones boost energy levels and augment memory, but over time they actually impair cognition, lead to depression, and alter the neural circuitry in

---

[91]Sherry Turkle, *Alone Together: Why We Expect More from Technology and Less from Each Other* (New York: Basic Books, 2011), 212; see also Jean M. Twenge, "Has the Smartphone Destroyed a Generation?," *The Atlantic* 320, no. 2, September 2017, 58-65; also Jean M. Twenge, *iGen: Why Today's Super-Connected Kids Are Growing Up Less Rebellious, More Tolerant, Less Happy—and Completely Unprepared for Adulthood—and What That Means for the Rest of Us* (New York: Atria, 2017).
[92]Turkle, *Alone Together*, 157.
[93]Ibid., 154.
[94]See Susan S. Phillips, *The Cultivated Life: From Ceaseless Striving to Receiving Joy* (Downers Grove, IL: InterVarsity Press, 2015).
[95]Michael Harris, *The End of Absence: Reclaiming What We've Lost in a World of Constant Connection* (New York: Penguin, 2014), 10.

the hippocampus, amygdala, and prefrontal cortex, the brain regions that control mood and thought.[96]

Chronic and prolonged "techno-brain burnout," as Small calls it, can eventually lead to permanent changes in the underlying structure of the brain.[97]

Interestingly, Harris also points out that computer and television screens prompt what is called an "orienting response," a basic brain function that redirects our attention.[98] This is why we find it so difficult *not* to glance at computer displays and television screens when they appear in our field of vision. Our brains have apparently evolved in such a way as to be ever ready to redirect our attention quickly from one thing to another. In a natural environment subject to rapid and potentially dangerous change, it is not difficult to see how the capacity for rapidly redirecting one's attention might improve one's chances of survival.

Yet here again, the orienting response can be over-stimulated. Researchers have found, for example, that the amount of time young children spend in of front of brightly illuminated screens—by some measures up to an average of ten hours per day—can interfere with brain development. Harris cites Douglas Gentile, a researcher at Iowa State University:

> We're now finding that babies who watch television in particular end up more likely to have attention deficit problems when they reach school age. It's pretty obvious: If you spend time with a flickering, flashing thing, it may leave the brain expecting that kind of stimulation.[99]

In short, researchers have found that whenever children are allowed to exceed the American Academy of Pediatrics recommended limit of one to two hours of recreational screen time per day, the incidence of attention deficit disorders rises significantly.[100] Reporting similarly on the future prospects of children

---

[96]Ibid.
[97]Ibid.
[98]Ibid., 120.
[99]Ibid., 121.
[100]Ibid; see also L. S. Pagani, C. Fitzpatrick, T. A. Barnett, and E. Dubow, "Prospective Associations Between Early Childhood Television Exposure and Academic, Psychosocial, and Physical Wellbeing by Middle Childhood," *Archives of Pediatrics and Adolescent Medicine* 164, no. 5 (2010): 425-31, doi:10.1001/archpediatrics.2010.50; Sylvain-Jacques Desjardins, "Toddlers and TV: Early Exposure Has Negative and Long-Term Impact," *Forum*, University of Montreal, May 3, 2010; D. A. Christakis, F. J. Zimmerman, D. L. Di Giuseppe, and C. A. McCarty, "Early Television Exposure and

accustomed to the Internet's uninterrupted connection to information and social media, one recent study suggested that these children may well display an increased facility for multitasking, but they will also probably exhibit "a thirst for instant gratification and quick fixes, a loss of patience, and a lack of deep thinking ability."[101]

That a good deal of time and money are even now being spent meticulously crafting websites that will successfully capture our attention is, needless to say, distressing. In a recent *Atlantic* exposé, Bianca Bosker draws attention to Stanford University's Persuasive Technology Lab, where "behavior designers" employ the insights of modern psychology to devise online experiences that are addictive. "McDonald's hooks us by appealing to our bodies' craving for certain flavors," Bosker writes. "Facebook, Instagram, and Twitter hook us by delivering what psychologists call 'variable rewards' [i.e., rewards delivered at random, which have proven to rapidly and strongly reinforce behavior]. Messages, photos, and 'likes' appear on no set schedule, so we check for them compulsively, never sure when we'll receive that dopamine-activating prize."[102] A recent best-selling book written by an alumnus of Stanford's behavioral design program is actually entitled *Hooked: How to Build Habit-Forming Products.*[103]

Here it has been noted that the internet has become an "attention economy" shaped almost entirely around the demands and requirements of advertising. Information and images are therefore presented in such a way as privilege our impulses over our conscious intentions, often appealing to sensuality, anger, outrage, and other strong emotional responses. As James Williams, the former Google strategist who built the metrics system for the company's global search advertising business, writes: "The attention economy incentivizes the design of technologies that grab out attention, [with the result that] [w]e've habituated ourselves

Subsequent Attentional Problems in Children," *Pediatrics* 113, no. 4 (2004), doi:10.1542/peds.113.4.708.

[101]Pew Internet/Elon University survey, cited in Catherine Steiner-Adair, with Teresa H. Barker, *The Big Disconnect: Protecting Childhood and Family Relationships in the Digital Age* (New York: HarperCollins, 2013), 58.

[102]Bianca Bosker, "Tristan Harris Believes Silicon Valley Is Addicting Us to Our Phones. He's Determined to Make it Stop," *The Atlantic*, November 2016, 56-65.

[103]Nir Eyal, with Ryan Hoover, *Hooked: How to Build Habit-Forming Products* (New York: Penguin, 2014).

into a perpetual cognitive style of outrage, by internalizing the dynamics of the medium."[104]

Recent findings relative to attachment disorders are similarly sobering. In respect to television screen time, for example, it has been observed that for every additional hour of television beyond the recommended limit a child watches at age twenty-nine months, the odds increase that the child will be more detached and unengaged as a fourth-grade student, less successful at math, heavier, and more likely to be bullied.[105] The lasting impact upon children when parents are distracted by devices is also becoming clear. For children to thrive they must be attended to, particularly by their parents. When parents are persistently distracted by devices, their children suffer for it.[106] We might imagine that the influence upon children of the transient content mediated by technological devices would pale in comparison with the deep, primal influence of their families. This is only true, however, to the extent that sustained, intimate, face-to-face communication is actually happening within their families.[107] "Research keeps coming back," writes psychologist Catherine Steiner-Adele, "to the role of parents in knowing their children, knowing what they need, and staying connected to our children IRL—in real life."[108] In her most recent work *Reclaiming Conversation: The Power of Talk in a Digital Age*, Sherry Turkle observers similarly:

> Whether a family chooses to create device-free "sacred spaces" at home or chooses to cultivate daily habits of family conversation—devices or no devices— children recognize a commitment to conversation. And they see it as a commitment to family and to them. I think this can make the difference between

---

[104]Williams quoted in Paul Lewis, "Our Minds Can Be Hijacked: The Tech Insiders Who Fear a Smartphone Dystopia," in *The Guardian* (October 6, 2017), www.theguardian.com/technology /2017/oct/05/smartphone-addiction-silicon-valley-dystopia.

[105]See Susan Pinker, *The Village Effect: How Face-to-Face Contact Can Make Us Healthier and Happier* (Toronto: Random House Canada, 2014), 167-68. See also Atushi Senju and Mark H. Johnson, "The Eye Contact Effect: Mechanisms and Development," in *Trends in Cognitive Sciences* 13:3 (January 3, 2009): 127-34; and Laura Pönkänen, Annamari Alhoniemi, Jukka M. Leppänen, et al., "Does It Make a Difference if I Have Eye Contact with You or with Your Picture? An ERP Study," Social Cognitive and Affective Neuroscience 6:4 (September 1, 2011): 486-94.

[106]Sometimes tragically. A recent YouTube post captured a child wandering into the path of a moving car while her mother, just a few steps behind her but distracted by her cell phone, failed to notice until after the child had been killed.

[107]Steiner-Adair, *The Big Disconnect*, 40.

[108]Ibid., 27.

children who struggle to express themselves and those who are fluent, between children who can reach out and form friendships and those who may find it hard to Find Their Friends.[109]

Turkle's research suggests, however, that "social media" and the pervasiveness of communications devices is making it more and more difficult for families to make just these kinds of choices and so to stay attached.

Yet beyond the disorders of attention and attachment, our use of and reliance upon technological devices appear to be responsible for a host of serious psychosocial conditions. In a disturbing book cleverly titled *iDisorder: Understanding Our Obsession with Technology and Overcoming Its Hold on Us,* research psychologist Larry Rosen catalogues a raft of studies that indicate the relation between our use of technology and numerous psychological disorders.[110] It has been found, for example, that excessive use of social media aggravates narcissistic personality disorders among young adults; that the use of wireless mobile devices exacerbates obsessive-compulsive disorders; that consumer technology frequently plays a prominent role in exacerbating an existing mood disorder—including bipolar disorder—and perhaps triggers the onset of such conditions; that the use of technology increases social isolation and reduces our ability to empathize with others; that technology aggravates hypochondria; that it worsens eating disorders; that it exacerbates voyeurism and the exploitation of women and children. "From being online and using social networking," Rosen writes, "to texting and instant messaging, to playing video games and listening to music, the more people engage in technology-related activities, the more they exhibit schizotypal [paranoid and/or delusional] behaviors."[111] Along similar lines, psychiatrist, doctor, and literary scholar Ian McGilchrist notes that a number of increasingly common psychopathological syndromes such as anorexia nervosa, multiple personality disorder, and autism are historically

---

[109]Sherry Turkle, *Reclaiming Conversation: The Power of Talk in a Digital Age* (New York: Penguin, 2015), 136.

[110]Larry Rosen, *iDisorder: Understanding Our Obsession with Technology and Overcoming Its Hold on Us* (New York: Palgrave MacMillan, 2012); Rosen's notes provide an extensive listing of recent research into the possible connections between the use of technology and a variety of mental illnesses.

[111]Ibid., 181.

quite new and apparently unique to modern technological civilization.[112] McGilchrist contends that there is little doubt that the development of mass technological culture—including urbanization, mechanization, and alienation from the natural world—has increased mental illness.[113] There is a large and growing body of evidence, in other words, to suggest that a good deal of modern machine technology is not necessarily—and perhaps not *at all*—conducive to human thriving.

Admittedly, the evidence runs counter to popular opinion, leading one researcher to comment:

> What I've been doing over the last few months is attempting to convince people that we're facing an issue that's as important and unprecedented as climate change. And I call it "mind change" because I think there are certain parallels. With mind change the issue is how new technologies, the new environment of two dimensions, might be impacting the human mind, and changing the way especially young generations may be thinking or feeling.[114]

Of course, determining just what accounts for "mind change" will require further research.[115] In the meantime, the issues surrounding our use of

---

[112]Ian McGilchrist, *The Master and His Emissary: The Divided Brain and the Making of the Western World* (New Haven, CT: Yale University Press, 2009), 404.

[113]Ibid., 407. McGilchrist fears that the fearful momentum of machine civilization may also be due to the fact that technology is being developed, managed, and administered by the kinds of people who are perhaps least sensitive to its dehumanizing effects. Machine civilization, he observes, exacerbates a number of mental illnesses, and in particular schizophrenia. Along this line, it has been observed that people with schizoid or schizotypal traits very often appear to be attracted to, and are often deemed particularly suitable for employment in, the very areas of science, technology, and administration that have become central within the technological milieu. "[A] culture with prominent 'schizoid' characteristics," he writes, "attracts to positions of influence individuals who will help it ever further down the same path. And the increasing domination of life by both technology and bureaucracy helps to erode the more integrative modes of attention to people and things which might help us to resist the advances of technology and bureaucracy, much as they erode the social and cultural structures that would have facilitated other ways of being, so that in this way they aid in their own replication" (408). Of course, McGilchrist touches on a very sensitive topic here. Yet it is surely the case that individuals with certain personality types seem to find it easier than others to work within the modern technological system and that this system often rewards them quite handsomely. That such people might go on to use their wealth and influence to further the interests of the system is simply to be expected.

[114]Rosen, *iDisorder*, 201; Rosen indicates that this quotation is from Andrew Maynard of the Risk Science Center, but the comments were actually made by Susan Greenfield in the video "Mind Change Is 'an Issue That's as Important and Unprecedented as Climate Change,'" *The Guardian*, August 15, 2011, www.theguardian.com/commentisfree/video/2011/aug/15/susan-greenfield-video.

[115]Current research has been reviewed by Henry H. Wilmer, Lauren E. Sherman, and Jason M Chein in "Smartphones and Cognition: A Review of Research Exploring the Links Between

technology will undoubtedly continue to be hotly debated. Suffice it here to say that there are very few early indications that modern technologies are enhancing human consciousness in any meaningful way. Instead, our technologies appear to be interfering with our ability—to use the familiar language of Martin Buber—to enter into I-Thou relationships with one another and with the larger world.[116]

Brave New World?

Of course, this litany could be lengthened. We might also have discussed the dehumanizing impact of cyber-bullying, cybercrime and other illicit uses of networked digital technologies, our tendency online to interact only with those who share our own views,[117] the rapid proliferation of "fake news" and the implications of disinformation for democracy,[118] the possibilities of cyberwarfare, the demise of humanities curricula in favor of STEM programs, and the like. But the point is not to exaggerate all of this, and neither is it to rubbish the many and remarkable benefits of modern machine technologies and systems. Nevertheless, everywhere we look today in our technological milieu, ordinary embodied human existence in the world appears to be at risk.

Admittedly, it is easy to overstate this risk and to exaggerate the threat that modern machine technology poses to human life. Perhaps this is because, as Thompson points out, we regard pessimism—in respect to our technological future—as more intelligent than optimism.[119] Cultural prophecies of doom in relation to new technologies, he reminds us, have occurred

---

Mobile Technology Habits and Cognitive Functioning," in *Frontiers in Psychology* (2017), 8:605, https://www.ncbi.nlm.nih.gov/pmc/articles/PMC5403814.

[116]See Martin Buber, *I and Thou*, trans. Walter Kaufmann (New York: Charles Scribner's Sons, 1970 [1937]).

[117]See Emerson T. Brooking and P. W. Singer, "War Goes Viral: How Social Media Is Being Weaponized," *The Atlantic*, November 2016, 70-83. "For all the hope that comes from connecting with new people and new ideas," Brooking and Singer write, "researchers have found that online behavior is dominated by 'homophily': a tendency to listen to and associate with people like yourself, and to exclude outsiders" (74).

[118]See, for example, Niall Ferguson, *The Square and the Tower: Networks and Power from the Freemasons to Facebook* (New York: Penguin, 2018).

[119]Clive Thompson, *Smarter Than You Think: How Technology Is Changing Our Minds for the Better* (New York: Penguin, 2013), 285-86; Thompson cites experiments conducted by Harvard's Teresa Amabile that indicate that when people hear negative, critical views, they regard them as inherently more intelligent than optimistic ones.

"with metronomic regularity, and in nearly identical form, for centuries."[120] With all new tools, the challenge is to become aware of their dangers and to negotiate how not to use them, while at the same time not becoming blind to technologies that "truly augment our thought and bring intellectual joy."[121]

Still, the evidence suggests that all is not well. While in many respects the modern technological milieu represents the acme of human genius and ingenuity, in many other respects it is strangely alien to human being. And as a great many modern commentators have warned, the emerging machine world appears to be one in which it will be difficult for many of us to live, much less to thrive. There seems to be a widening discrepancy between our collective technological accomplishments and the subjective condition of individual human persons. Chesterton's "huge modern heresy" again comes to mind in this connection. Rather than serving us, our machine systems appear to require us to adapt and to serve them.

Of course, it may well be the case that those occupying privileged positions on the leading edges of technological development are enjoying new possibilities, and—who knows?—perhaps even an expansion of consciousness. Still, most of the rest of us seem to be suffering a kind of diminishment. Lewis Mumford already lamented in 1934,

> Too dull to think, people might read: too tired to read, they might look at the moving pictures: unable to visit the picture theater they might turn on the radio: in any case, they might avoid the call to action: surrogate lovers, surrogate heroes and heroines, surrogate wealth filled their debilitated and impoverished lives and carried the perfume of unreality into their dwellings. And as the machine itself became, as it were, more active and human, reproducing the organic properties of eye and ear, the human beings who employed the machine as a mode of escape have tended to become more passive and mechanical. Unsure of their own voices, unable to hold a tune, they carry a phonograph or a radio set with them even on a picnic: afraid to be alone with their own thoughts, afraid to confront the blankness and inertia of their own minds, they turn on the radio and eat and talk and sleep to the accompaniment of a continuous stimulus from the outside world: now a band, now a bit of propaganda, now a piece of public gossip called news. Even such autonomy as the poorest

---

[120]Ibid.
[121]Ibid., 288.

drudge once had, left like Cinderella to her dreams of Prince Charming when her sisters went off to the ball, is gone in this mechanical environment: whatever compensations her present-day counterpart may have, it must come through the machine. Using the machine alone to escape from the machine, our mechanized populations have jumped from a hot frying pan into a hotter fire.[122]

Remarkably, Mumford's mid-twentieth-century reflections concerned the impact of the machine upon nineteenth-century culture. Needless to say, he would likely not be impressed with where we have come since then.

So Why Don't We Change It?

Why hasn't there arisen more social pressure to redress the "huge modern heresy"? This is an interesting question, and we will spend a good deal of time trying to answer it in the next two chapters. Plainly, the adverse consequences of machine technology have hitherto been offset by the material benefits of technologically driven economic growth. This growth has fueled a steady rise in the standard of living for generations, and perhaps it will continue to do so for some time to come. No one wants to upset, much less kill the goose that has laid so many golden eggs for such a long time.

We have undoubtedly also been enticed by what Borgmann has called "the promise of technology" to liberate us from want and drudgery. Along this line, we routinely assume that the pre-technological past must have been much worse and that the lion's share of "human progress," which, since the eighteenth century, has been largely defined in terms of material welfare, has been made possible by science and technology. It is ironic, then, that so much evidence suggests that our technologies have actually led—for the most part unintentionally—to distraction and disengagement from the larger world as well as from each other.

Our inability to accurately assess the costs and benefits of modern machine technology probably also has to do what Carr identifies as the "paradox of work."[123] Carr notes that, according to recent psychological studies, we are simply not very good at identifying in advance which sorts of things will actually satisfy us and which will leave us discontented. Indeed, we are often

---

[122]Mumford, *Technics and Civilization*, 315-16.
[123]Carr, *The Glass Cage*, 15.

inclined to desire the very things that will not, in the end, yield satisfaction and fulfillment. We are also inclined to assume that we will enjoy things that come easily, and that leisure is to be preferred over labor. It is no wonder, then, that we are enamored with automated machinery. As Carr observes, "By offering to reduce the amount of work we have to do, by promising to imbue our lives with greater ease, comfort, and convenience, computers and other labor-saving technologies appeal to our eager but misguided desire for release from what we perceive as toil." In consequence, we simply aren't very good at evaluating the costs and benefits of machine technology. "The deck is stacked," Carr concludes, "economically and emotionally, in automation's favor."[124]

The Cash Nexus

Maybe we just need to give modern machine technology more time to develop. Perhaps the sorts of problems that we have identified stem from the fact that our technologies are relatively new and have yet to disclose their full potential for enhancing human consciousness. Although not impossible, there are good reasons to suspect that such an outcome is not very likely. As Borgmann has noted, whatever newer and better technologies lie just around the corner, they will still be made available to us as commodities to be purchased and consumed. As such, they will probably not demand too much from us in the way of commitment, discipline, and/or skill.

And it is here that we touch on what is perhaps the most significant reason that there has not been more resistance to modern technological development: it is so closely bound up with economic interests. Indeed, while we have discussed the historical importance of the movement from tools and contrivances to scientifically derived machine technology, it was—and is— the powerful synthesis of machine technology with capitalism that explains the unique potency of modern industrial civilization. Our remarkable technical skill, in other words, isn't the only thing that stands in the way of creative and constructive thinking about modern automatic machine technology. We are also going to need to grasp and to reckon with the combination of technological innovation and creative business organization that has been, and

---

[124]Ibid., 17.

continues to be, the source of Western prosperity. In short, it is the insidious logic of money—itself an example of an ingenious premodern technology—that also impedes our ability to "comprehend technique from beyond its own dynamism." This, then, is what we will examine in our next chapter.

CHAPTER TWO

THE MOMENTUM AND INERTIA OF MODERN TECHNOLOGICAL DEVELOPMENT

To discuss the proliferation of inventions during the last two centuries . . .
without reference to this immense pecuniary pressure constantly exerted
in every technological area, is to ignore the most essential clue to the
seemingly automatic and uncontrollable dynamism of the whole system.

LEWIS MUMFORD, *THE MYTH OF THE MACHINE*

THE WORD *MOMENTUM* DENOTES the driving power, strength, and/or speed of an object in motion. *Inertia* means that an object will remain in motion—moving in a given direction and at a particular speed—so long as it is not acted upon by an external force. Both terms are useful in describing modern technological development. Modern machine technology has picked up a good deal of momentum over the last 150 years, particularly since the late 1950s, as it has been augmented by digital information and communications technologies. Its inertia, furthermore, is plainly in the direction of automatism.

My concern, of course, is that automatic machine technology is pushing society and culture away from ordinary embodied human existence, and at considerable speed. Our technologies are forcing us to become more and more machinelike, in spite of the physiological, psychological, and social toll this appears to be taking on us. At the extreme, our technology seems bent on simply getting rid of us. Overcoming its momentum and inertia is going to require concerted efforts.

As we have seen, the modern "factory system" and its offspring, today's production and distribution facilities, were from the beginning designed to eliminate the human as much as possible from the machine process. Ironically, it appears that we are being diminished by the many of the same technologies that promise to make our lives easier, safer, more convenient, and more pleasurable. Even "tech" insiders are beginning to voice fairly serious reservations about the momentum and direction of modern technological development.[1]

Yet even the critics of modern technology tend to underestimate its inertia and the external forces that would be required to divert and/or redirect it. Jaron Lanier, for example, a pioneer in the field of virtual reality, believes that young computer scientists would surely keep human beings at the center of new designs if only they would periodically remind themselves to do so.[2] He suggests that if we could just be made to realize how egregiously large firms are taking advantage of our personal data, we could reform the system along more humanistic lines. The spiritual challenge, Lanier avers, is simply to keep from losing touch with "that core of experience, that little something that doesn't fit into the aspects of reality that can be digitized."[3]

Unfortunately, the inertia of modern technological development is not likely to be disturbed by such insipid cautions. Modern machine technology has been running roughshod over "that little something" at the core of human experience for some time now, with great success and largely without compunction. Obviously, a more focused effort is going to be required if we want to redirect modern machine technology in a more human direction. The purpose of this second chapter, then, is to begin to show just how much more is going to be required of us. Our strategy is to clarify how modern technological development dovetails with powerful economic forces and interests. Modern technology's momentum and inertia stem, in very large

---

[1]See Jaron Lanier, *Who Owns the Future?* (New York: Simon & Schuster, 2013); also Andrew Keen, *The Internet Is Not the Answer* (New York: Atlantic Monthly Press, 2015).
[2]Lanier, *Who Owns the Future?*, 362.
[3]Ibid., 364.

part, from the fact that our technologies have emerged out of the matrix of modern capitalism. We must understand this linkage if we are to stand any chance of reforming and/or redirecting the modern system.

The Economic Context of the Modern Technological Takeoff

Modern machine technology emerged in Western Europe after the seventeenth century. But why then? And why there? After all, early-modern Europe—recently devastated by religious warfare—was hardly the most likely setting for such a development. China and the Arab world were both more technologically advanced in the seventeenth century, and so a hypothetical observer familiar with the world's civilizations at the time likely would not have chosen Europe as the site of the modern technological takeoff.

Yet a number of factors in seventeenth-century Western Europe coalesced to form a climate favorable to technological innovation. Mumford grouped these factors beneath three broad headings. Under the first, "Cultural Preparations," he discussed the importance of theological, aesthetic, and military influences, as well as the importance of new conceptions of space and time. "Men had become mechanical," Mumford contended, "before they perfected complicated machines to express their new bent and interest; and the will-to-order had appeared once more in the monastery and the army and the counting-house before it finally manifested itself in the factory."[4]

Mumford went on to stress that what lay behind the remarkable technological innovations of the eighteenth and nineteenth centuries in Europe was not simply a string of technical advancements and improvements but rather a change of *mind*. "Before the new industrial processes could take hold on a great scale," he wrote, "a reorientation of wishes, habits, ideas, goals was necessary."[5] Mumford sought to capture this new change of mind in the expression "the mechanical world-picture,"[6] and he suggested that a new "mechanical" outlook was every bit as important to the development of modern technics as were subsequent scientific discoveries and technical inventions.

---

[4]Lewis Mumford, *Technics and Civilization* (London: Routledge & Sons, 1934), 3.
[5]Ibid.
[6]Ibid., 25.

Mumford also underscored the importance of early-modern science. For modern science turned Western attention toward quantities, "objects" that could be counted, weighed, and measured, as well as toward time sequences that could be repeated and predicted. Science also sought data that could be objectified and analyzed systematically, eliminating the subjectivity of the human observer as much as possible. The truths thus revealed by scientific procedures were to be true for anyone at any time and in any place.

Science was also concerned to reduce complexity to simplicity by breaking down natural objects and processes into their component parts, parts that could be further subdivided until nature's basic components and processes were revealed. Science thus came to conceive of nature as an elaborate mechanism, a kind of intricate machine, often described by the analogy of clockwork. This new scientific conception of nature was undoubtedly crucial to the emergence and development of modern machine technology.

Yet as important as science was for the emergence of modern technics, Mumford was especially concerned to highlight the importance of *capitalism* to modern technological development. For the market system provided the context within which machine technology could develop and thrive. Not only did capitalism create powerful incentives for making mechanical improvements, it also provided the capital necessary for experimentation and for bringing these improvements quite literally to market.[7] Without the market system's direction of resources to mechanical inventions, the mechanical world picture and the early-modern scientific outlook might simply have remained elite pastimes, interesting only to the few who enjoyed the leisure to delve into them. It was capitalism that connected the new mechanical orientation to the lives and concerns of ordinary people.

In view of the diminishment of ordinary, embodied human being, furthermore, the importance of capitalism to modern technological development can hardly be overstated. It is not simply that the intrinsic logic of modern machinery tends in the direction of automatism—for of course it does—but even more importantly the cost-effectiveness of machine versus

---

[7]Ibid., 27.

human labor has exerted constant pressure on modern designers and engineers to eliminate the human presence as much as possible from the machine process.

Yet here we come full circle. For one of the keys to understanding modern capitalism is the unique use it makes of that ancient but common technology we call *money*. At a particular juncture in Western history, capitalism's unique use of money apparently became linked to the mechanical world picture of early modern science. This linkage precipitated a new and powerful industrial paradigm, a paradigm that continues to shape our world. In what follows I want to consider this momentous development first in terms of the intrinsic logic of money, and second—historically—by taking a closer look at just when and how capitalism's use of money became intertwined with the insights of early modern science. In this light, we will consider the implications of the close connection between technics and economics for thinking about contemporary technological development.

The Logic of Money

Money is a token of value. It may be defined as any clearly identifiable object or substance generally accepted as payment for goods and services, as well as for the repayment of debt within a given market or jurisdiction. Money facilitates exchange relationships as well as serving as an aid to memory. It enables us to reckon, to store, and to transmit value, and in doing so it confers both power and security. It is a truly ingenious device.

Money is an ancient technology. Historians believe that it developed out of the barter economy when durable and easily transportable objects of various kinds were made to serve as intermediate stores of value for goods that were, at some future date, to be exchanged. Simple and local tokens eventually came to be more widely recognized, developing into "commodity money" of standardized value that enhanced money's usefulness as well as its reach. As Jack Weatherford notes,

> Once human technology and social organization developed to the point of using standardized amounts of gold and silver in exchange, it became only a matter of time before smaller coins appeared. The technological and cultural leap from primitive coins constituted the first money revolution in history,

and to the best of numismatic knowledge, it happened only once. It took place in western Asia in what is today Turkey [apparently in Lydia ca. 640 BC], and from there it spread around the world to become the global money system and the ancestor of the system in which we live and work today.[8]

Money makes all kinds of things, including thinking, easier and more convenient. It enables us to objectify our diverse and varied experiences and to unify them into a system of calculable "values," which is to say, into quantities that can then be added, subtracted, multiplied, and divided. Money thus encourages quantitative and abstract thought.[9] The use of money is thought to have been crucial to the development of mathematics and, by extension, to what would eventually become scientific understanding.

Much like phonetic literacy, then, money is a technology that transformed human consciousness. "Humans have found many ways to bring order to the phenomenological flow of existence," writes Weatherford in this connection. He continues:

> Money is one of the most important. Money is strictly a human invention in that it is itself a metaphor; it stands for something else. It allows humans to structure life in incredibly complex ways that were not available to them before the invention of money. This metaphorical quality gives it a focal role in the organization of meaning in life. Money represents an infinitely expandable way of structuring value and social relationships—personal, political, and religious as well as commercial and economic.[10]

But of course, this infinitely expandable way of structuring value and social relationships, based as it is upon money's conversion of qualities into mere quantities, also poses a profound threat to human cultures—for it acts as a kind of universal solvent, dissolving everything into numbers. Karl Marx captured this vividly in a famous passage in *Capital* (ca. 1870), cited by Mumford:

> Since money does not disclose what has been transformed into it, everything, whether a commodity or not, is convertible into gold. Everything becomes

---

[8]Jack Weatherford, *The History of Money: From Sandstone to Cyberspace* (New York: Three Rivers Press, 1997), 27.
[9]Ibid., 148.
[10]Ibid., 43.

salable and purchasable. Circulation is the great social retort into which everything is thrown and out of which everything is recovered as crystallized money. Not even the bones of the saints are able to withstand this alchemy; and still less able to withstand it are more delicate things, sacrosanct things which are outside the commercial traffic of men. Just as all qualitative differences between commodities are effaced in money, so money, a radical leveler, effaces all distinctions.[11]

Expanding upon Marx's insights, sociologist Georg Simmel observed that our use of money and monetary calculations to store value can actually have the effect of flattening our experience of the world, its colorlessness rendering us indifferent to individual qualities: "Irreparably it hollows out the core of things," Simmel wrote, "their individuality, their specific value, and their incomparability. All things float with equal specific gravity in the constantly moving stream of money."[12]

Yet precisely because it is so useful, conferring both power and security, money can become a kind of fetish or even the object of worship. Traditional religions have tended to be deeply suspicious of "Mammon" and have placed all sorts of restrictions around its use. Biblical religion, for example, condemned the charging of interest on most loans as a kind of taking advantage of one's neighbor. "No one," Jesus warned his followers, "can serve two masters. Either you will hate the one and love the other, or you will be devoted to the one and despise the other. You cannot serve both God and money" (Mt 6:24). The love of money, the apostle Paul insisted, "is a root of all kinds of evils" (1 Tim 6:10).

Religious misgivings notwithstanding, money's impact upon human cultures has been far-reaching. Nowhere has this impact been more keenly felt than within modern technological/industrial civilization, where modern capitalism has put money to new and particularly productive uses.[13] In this connection, we note that Max Weber famously defined capitalistic economic

---

[11]Karl Marx, cited in Mumford, *Technics and Civilization*, 23-24; originally from Karl Marx, *Capital: A Critique of Political Economy*, ed. Frederich Engels (New York: The Modern Library, 1906), 148.

[12]Georg Simmel, "The Metropolis and Mental Life," in *The Sociology of Georg Simmel*, trans. Kurt H. Wolff (New York: Free Press, 1950), 414.

[13]For a somewhat more detailed discussion of capitalism's use of money, see Craig M. Gay, *Cash Values: Money and the Erosion of Meaning in Today's Society* (Grand Rapids: Eerdmans, 2004).

action as that "which rests on the expectation of profit by the utilization of opportunities for exchange."[14] The English word *profit* comes from the Latin *profectus*, which simply meant "to advance." We commonly use the word to refer to the excess of returns over expenditures following a series of transactions, something that is determined by careful accounting. Along this line, Weber insisted that the development of rational accounting, and in particular the invention of double-entry bookkeeping, was crucial to the development of early capitalism. These clever techniques enabled managers to assess the effectiveness of their economic activities with great precision. Accounting techniques and procedures depended, in turn, upon the prior objectification of value that had already been made possible by the conventional use of money.

As Weber's definition suggests, capitalistic economic action doesn't simply recognize profit retrospectively. Rather, it also plans to utilize opportunities for exchange on the expectation of future profits. As economic historian Joseph Schumpeter observed, modern businessmen have learned to utilize increasingly sophisticated techniques and procedures, effectively turning money's crystallization of value into a tool for forecasting growth.

This has powerfully propelled the logic of modern enterprise.[15] Market research, cost-benefit analyses, economic forecasting, and a host of other techniques—all premised upon money's precise numerical restatement of value—are now employed by producers to determine whether production should be scaled up or down; by managers in mitigating risk and in attempting to steer their organizations toward opportunities; by investors in seeking to maximize return on investment; and by all the rest of us as we seek to get the biggest "bang for our buck."

The role of money in modern industrial capitalist society has thus become, as Simmel observed, "absolute for the consciousness of value,"[16] in a manner that could not have been conceived by our ancestors. As values came to be

---

[14]Max Weber, *The Protestant Ethic and the Spirit of Capitalism*, trans. Talcott Parsons (New York: Charles Scribner's Sons, 1958), 17.

[15]Joseph A. Schumpeter, *Capitalism, Socialism and Democracy* (New York: Harper & Row, 1975 [1942]), 123.

[16]Georg Simmel, *The Philosophy of Money*, trans. David Frisby (London: Routledge & Kegan Paul, 1978), 250.

objectified and quantified for the economic sector, Schumpeter observes, "this type of logic or attitude or method then starts upon its conqueror's career subjugating—rationalizing—man's tools and philosophies, his medical practice, his picture of the cosmos, his outlook on life, everything in fact including his concepts of beauty and justice and his spiritual ambitions."[17] In short, the influence of money pervades almost all of modern life.

Just as automatic machine technology has moved modern society and culture in the direction of objectification, rationalization, and automation, furthermore, so that the ancient technology we ordinarily call money, in addition to the techniques and procedures we call accounting, have moved us in very similar directions, particularly as these techniques and procedures were combined with the outlook and insights of early modern science. The market system's continuous coordination of buyers and sellers by means of prices, it has been said, was one of the first truly automatic modern machines.

Cost/Profit Calculation + Experimental Science

Beginning in the seventeenth century, enterprising capitalists took notice of the insights that were being generated by early-modern science's new understanding of nature. In relatively short order, cost-profit calculations appear to have been conjoined with early-modern science in an immensely potent synthesis that would lead within a century to what is commonly called the Industrial Revolution, an economic and technological upheaval that continues to shape our world.

But how could this have happened so quickly? After all, human cultures have historically been deeply suspicious of both profit and innovation and have placed many restrictions on them. Innovation always benefits some more than others and thus inevitably disrupts what is typically a delicately balanced status quo: within any zero-sum economic system, one person's profit must necessarily come at someone else's expense, hence the Old Testament prohibition of lending money at interest. Again, late medieval Christian civilization, while economically and technologically dynamic in many respects, was not where and when one would have expected the

---

[17]Schumpeter, *Capitalism, Socialism and Democracy*, 124.

modern industrial system to have taken off. As Nathan Rosenberg and L. E. Birdzell Jr. comment,

> We can hardly grasp the degree to which it was alien, in the Middle Ages, for anyone to guide present economic activity by the deliberate calculation of future consequences. In both country and town, what one did for a living was what one had done for years past and expected to continue to do, in the same way and on the same terms, until death ended the round of sowing and reaping. . . . The very idea of varying and changing what one did in response to calculations of future consequences and present conditions of supply and demand lay outside the normal pattern of medieval economic life. . . . The possibility of calculation, of assessing prospective magnitudes of cost and revenue and the probability of alternative outcomes in a novel enterprise, of profiting from judicious buying and selling (how could it be done if both prices were "just"?) rather than from diligent service to one's lord or from industriously plying one's trade, was wholly alien to the customary order of feudal society.[18]

As such observations suggest, social and economic life for most people, and for most of human history, has been disciplined by restrictions upon profit seeking and innovation. The burden of justification has fallen fairly heavily upon those who advocate any potentially disruptive change.

This was almost certainly the case in Europe at the beginning of the modern period. Economic historian Robert Heilbroner observes this in his classic introduction to economic thought, *The Worldly Philosophers*, by relating Colbert's 1666 legislation designed to squelch innovation in the weaving industry. The French Comptroller General decreed, among other things, that "the fabrics of Dijon and Selangey are to contain 1,408 threads including selvages, neither more nor less."[19] At least by the middle of the seventeenth century in France, Heilbroner noted, the legitimacy of trying to profit by means of innovation, technological or otherwise, had not yet taken root.[20]

Yet for a variety of reasons, Europe's traditional restrictions on profit and innovation began—probably even as Colbert's legislation was being

---

[18]Nathan Rosenberg and L. E. Birdzell Jr., *How the West Grew Rich: The Economic Transformation of the Industrial World* (New York: Basic Books, 1986), 52-53.
[19]Robert Heilbroner, *The Worldly Philosophers: The Lives, Times, and Ideas of the Great Economic Thinkers*, 7th ed. (New York: Simon & Schuster, 1995), 23.
[20]Ibid., 24.

introduced—to dissolve. By the eighteenth century, such restrictions had all but disappeared from European society and culture. Just why they disappeared is one of the questions Weber sought to answer in his celebrated analysis of the origins of the modern market economy, *The Protestant Ethic and the Spirit of Capitalism*.[21]

Capitalism, Weber wrote, "is identical with the pursuit of profit, and forever renewed profit, by means of continuous, rational, capitalistic enterprise."[22] Profit, as we have seen, is a technical term that refers, not to personal gain per se, but rather to successful economic action. Indeed, forever renewed profit all but requires that most of the gains attained by way of successful economic activity be reinvested in one's enterprise and not personally consumed. In this connection, Weber noted that the "pursuit of forever renewed profit by means of continuous, rational, capitalistic enterprise" is historically peculiar.[23] Not only does it require rational accounting procedures, the separation of businesses from households, and the rational and reliable administration of law, but it must also be animated by a particular spirit or ethic that is industrious and yet also frugal and ascetic, which is to say, world-denying. For Weber, this new spirit was epitomized in the folk aphorisms offered in Benjamin Franklin's *Poor Richard's Almanac* (ca. 1735), for example, that time is money, that credit is money, and that money is of a prolific, generating nature.[24] The *summum bonum* of this new ethic, Weber stressed, seems to have been "the earning of more and more money, combined with the strict avoidance of all spontaneous enjoyment of life."[25] Weber wrote:

> Truly what is here preached is not simply a means of making one's way in the world, but a peculiar ethic. The infraction of its rules is treated not as foolishness but as forgetfulness of duty. That is the essence of the matter. It is not mere business astuteness, that sort of thing is common enough, it is an ethos. *This* is the quality which interests us.[26]

---

[21]Weber, *Protestant Ethic*; see also my discussion of the Weber thesis in *Cash Values*, 26-31.
[22]Weber, *Protestant Ethic*, 17.
[23]Ibid.
[24]Ibid., 48.
[25]Ibid., 53.
[26]Ibid., 51 (emphasis in original).

Of course, the period during which this peculiar ethos actually animated the behavior of early-modern entrepreneurs was relatively brief, and Weber was quick to add that present-day capitalism no longer requires the conscious acceptance of such ethical maxims. Still, such an acquisitive-*cum*-ascetic ethos does appear to have been necessary to set the capitalist system in motion.

Where did this peculiar ethos come from? Weber's answer, of course, was that it could be traced to Protestant Christianity, and particularly to Calvinism. Basically, Martin Luther's repudiation of the medieval distinction between "sacred" and "secular" callings[27] meant that Protestants were more comfortable in affirming everyday work in the world than were their Roman Catholic counterparts, who tended (at least theologically speaking) to view work as a kind of necessary evil. In contrast, Protestants insisted that all practical activity in the world—or at least all practical activity not at odds with the moral law—was not only religiously legitimate, but was to be considered an essential aspect of Christian obedience to the divine calling, or vocation. Calvinists appear to have unleashed the radical social potential of this Protestant understanding of vocation by insisting that the Christian is called, not just to work in the world, but to the careful, deliberate, and energetic reform of the entire social order. Every aspect of it was to be brought into accord with God's revealed will. "It may seem that Reformed social ethics favours preservation of a static social order," André Biéler writes in a treatise on Calvin's social and economic thought. "Not so," he continues. "On the contrary, the gospel is a dynamic force that always tends to reform the established order."[28]

Much of this reform was to be accomplished by means of basic practical and instrumental rationality. This, Weber believed, explains the curious admixture of otherworldliness and utilitarianism in Calvinist ethics:

> It seems at first a mystery how the undoubted superiority of Calvinism in social organization can be connected with this tendency to tear the individual away from the closed [traditional social] ties with which he is bound to this

---

[27]See Martin Luther, "To the Christian Nobility of the German Nation," trans. Charles M. Jacobs, rev. James Atkinson, in *Martin Luther: Three Treatises* (Philadelphia: Fortress, 1970), 12-18.
[28]André Biéler, *Calvin's Economic and Social Thought*, ed. Edward Dommen, trans. James Greig (Geneva: WCC Publications, 2005), 231.

world. . . . But God requires social achievement of the Christian because He wills that social life shall be organized according to His commandments, in accordance with that purpose. . . . For the wonderfully purposeful organization and arrangement of this cosmos is, according to both the revelation of the Bible and to natural intuition, evidently designed by God to serve the utility of the human race. This makes labour in the service of impersonal social usefulness appear to promote the glory of God and hence to be willed by him.[29]

Calvinists, in short, tended to be more practical and pragmatic than either their Roman Catholic or their Lutheran counterparts in assessing the results of economic and technological undertakings. If something worked and proved useful, then for that reason it was deemed to be Christianly acceptable. The Christian was not to be bound by moribund tradition in such matters, but instead was expected to be open to practical-rational innovation in the service of God and neighbor and in the interests of the common good. Puritan theologian Richard Baxter's advice is commonly cited in this connection:

> "If God show you a way in which you may lawfully get more than in another way (without wrong to your soul or to any other)," Baxter insisted, "if you refuse this, and choose the less gainful way, you cross one of the ends of your calling, and you refuse to be God's steward, and to accept His gifts and use them for Him when He requireth it: you may labour to be rich for God, though not for the flesh and sin."[30]

Within this new Protestant ethic, believers were all but mandated to improve the material conditions of life by inventing new devices, increasing productivity, diminishing the costs of production, hastening transport, facilitating communications, and opening up new markets for distribution. This ethic was naturally open to the practical empiricism of early-modern science. It found particularly fertile soil in the "New World" of North America, where it was brought by various refugee groups— mostly Protestants—beginning in the seventeenth century. Its peculiar energy made a strong impression on Tocqueville during his visit to America in the early nineteenth century. "Most of the people in these [democratic] nations," he wrote (ca. 1830),

---

[29]Weber, *Protestant Ethic*, 108.
[30]Baxter cited in Weber, *Op. cit.*, 162.

are extremely eager in the pursuit of immediate material pleasures and are always discontented with the position they occupy and always free to leave it. They think about nothing but ways of changing their lot and bettering it. For people in this frame of mind every new way of getting wealth more quickly, every machine which lessens work, every means of diminishing the costs of production, every invention which makes pleasures easier or greater, seems the most magnificent accomplishment of the human mind.[31]

Canadian political philosopher George Grant underscored the world-historical significance of the synthesis of Calvinism and the new empirical and scientific spirit in an intriguing essay entitled "In Defense of North America."[32] Grant noted that both theologians and early-modern scientists had sought to escape from the constraints and errors of traditional medieval understanding, but for different reasons. The scientists simply wanted to be allowed to observe nature unencumbered by Aristotelian teleology, and hence to see the world empirically, which is to say, as it actually *is*. Protestant theologians attacked the medieval teleological doctrine because it led people away from fundamental reliance upon the Bible. Both groups were therefore open to exploring new ways of getting things done in the world. Calvinism's distinctive understanding of God, Grant noted, left the settlers of the North American continent particularly open to empiricism and utilitarianism. For our purposes, this is to say that it left them particularly open to profit and innovation. Grant cited social theorist Ernst Troeltsch in this connection:

> Calvinism, with its abolition of the absolute goodness and rationality of the Divine activity [in creation] into mere separate will-acts [i.e., "decrees"], connected by no inner necessity and no metaphysical unity of substance, essentially tends to the emphasizing of the individual and empirical, the renunciation of the conceptions of absolute causality and unity, the practically free and utilitarian individual judgment of all things. The influence of this spirit is quite unmistakably the most important cause of the empirical and positivist tendencies of the Anglo-Saxon spirit.[33]

---

[31]Alexis de Tocqueville, *Democracy in America*, trans. George Lawrence (Garden City, NY: Doubleday, 1969), 462.
[32]George Grant, "In Defense of North America," in *Technology and Empire: Perspectives on North America* (Toronto: Anansi, 1969), 15-40.
[33]Ibid., 21.

Of course, Troeltsch (and Weber too, for that matter) has been criticized for misunderstanding and misrepresenting Calvinist theology, and we will have occasion to address the ideational origins of the modern technological outlook in greater detail in our next chapter. Yet whatever the reasons for the collapse of traditional restrictions upon profit seeking and innovation, both have, often in combination, come to be encouraged and even celebrated in the modern West.

In what we commonly call the market system, profit seeking is believed to serve profoundly positive social and political purposes. Echoing Adam Smith, Heilbroner characterizes the modern market system as an "astonishing arrangement in which society assured its own continuance by allowing each individual to do exactly as he saw fit—provided he followed a central guiding rule ... [i.e., that] each should do what was to his best monetary advantage."[34] "In the market system," he continues, "the lure of gain, not the pull of tradition or the whip of authority, steered the great majority to his (or her) task. And yet, although each was free to go wherever his acquisitive nose directed him, the interplay of one person against another resulted in the necessary tasks of society getting done."[35] Why? Because the market's remarkable coordination of thousands of individual decisions made by tens of thousands of individual buyers and sellers directed—by means of Smith's celebrated "Invisible Hand"—capital and human effort toward those things that society needed done and away from those that society didn't need done. This, Heilbroner insists, "was the most important revolution, from the point of view of shaping modern society, that ever took place—fundamentally more disturbing by far than the French, the American, or even the Russian Revolution."[36]

Rosenberg and Birdzell contend similarly that the West has been able to grow and thrive precisely because political pluralism and the relative flexibility of Western institutional life created a social space within which profit seeking and practical experimentation could develop and flourish together. Indeed, they conclude that the West has grown rich by combining creative

---

[34]Heilbroner, *Worldly Philosophers*, 20.
[35]Ibid., 20-21.
[36]Ibid., 21.

business organization with technological innovation "to harness resources to the satisfaction of human wants."[37] They label this potent synthesis of experimental technology and business organization the West's "growth system,"[38] and they suggest that its key features were (and still are) that it has distributed the authority and resources necessary for innovation quite broadly; that it has lessened if not altogether eliminated political and religious restrictions upon innovation; and that it permits inventors and investors to reap the financial rewards of their inventions and/or their investment in innovation.

The West's "growth system" quickly yielded innovations in trade and commerce, in the discovery of new resources, in production techniques, and in business organization and management. Citing mining and woodcraft, Mumford pointed out how one kind of innovation seemed naturally to have led to another.[39] Increasingly deep mines required new and more effective kinds of machinery (particularly mechanical pumping equipment). The new mines also required new sources of capital to pay for this increasingly expensive machinery. This, in turn, led to new forms of ownership of the mines, which led to new relations between miners and the now largely absentee mine owners. Serendipitously, early modern mining supplied the metals necessary for mechanical experimentation as well as for nascent industrial development. And, of course, the mines supplied the metals necessary for the expansion of a "sound but expansible currency."[40] Lastly, Mumford noted that reckless speculation in mining stocks led to new developments in law and finance aimed at mitigating risk.[41]

Mumford made similar observations about the significance of early modern woodcraft to what would become industrial development. It was the woodworkers, after all, who developed all manner of wheels as well as what Mumford considered to be the greatest of all machine tools, the lathe. "If the boat and the cart are the woodman's supreme contribution to transport," he observed, "the barrel, with its skillful use of compression and tension to

---

[37]Rosenberg and Birdzell, *How the West Grew Rich*, 33.
[38]Ibid., 20.
[39]Mumford, *Technics and Civilization*, 74.
[40]Ibid., 75.
[41]Ibid.

achieve water-tightness is one of his most ingenious utensils: a great advance in strength and lightness over clay containers."[42]

Yet Mumford was quick to add that the chief propagator of the machine in early-modern Europe was undoubtedly warfare.[43] Guns in particular required high-quality metals and accurate technologies for machining standardized parts to close tolerances.[44] Already by the end of the eighteenth century, massive armies had led to military mass production.[45] Mumford wrote:

> With an army of 100,000 soldiers, such as Louis XIV had, the need for uniforms made no small demand upon industry: it was in fact the first large-scale demand for absolutely standardized goods. Individual taste, individual judgment, individual needs, other than dimensions of the body, played no part in this new department of production: the conditions for complete mechanization were present.[46]

Warfare also quickly consumed much of what had been produced, creating the demand for further production.[47]

In his classic study of the Industrial Revolution in Britain, historian T. S. Ashton emphasized the close relationship between business interests and technological innovation:

> The industrial revolution was an affair of economics as well as of technology: it consisted of changes in the volume and distribution of resources, no less than in the methods by which these resources were directed to specific ends. The two movements were, indeed, closely connected. Without the inventions industry might have continued its slow-footed progress—firms becoming larger, trade more widespread, division of labor more minute, and transport and finance more specialized and efficient—but there would have been no industrial revolution. On the other hand, without the new resources the inventions could hardly have been made, and could never have been applied on any but a limited scale. It was the growth of savings, and of a readiness to put these at the disposal of industry, that made it possible for Britain to reap the harvest of her ingenuity.[48]

---

[42]Ibid., 80.
[43]Ibid., 85.
[44]Ibid., 87.
[45]Ibid., 90.
[46]Ibid., 92.
[47]Ibid., 94.
[48]T. S. Ashton, *The Industrial Revolution, 1760–1830* (Oxford: Oxford Paperbacks, 1969), 66.

The Industrial Revolution was not simply a revolution in scale and quantity, resulting in the production of more and more of the same old things. It was also a revolution in variety, resulting in the appearance of newer, better, and entirely different things. The significance of novelty within the modern system probably cannot be overstated. It was famously captured by Schumpeter in the oxymoron "creative destruction."[49] The disruptive nature of the capitalist economy, Schumpeter believed, is not simply due to the fact that economic life responds to and in a sense mirrors other sociological changes, such as wars and revolutions. "Such things," Schumpeter contended, "do condition industrial change, but they are not its prime movers."[50] Nor is the disruptive character of capitalism due primarily to population growth and/or to changes in monetary policies and/or capital flows. No, what explains the restless character of modern capitalism is the system's incessant need to *grow*, and hence its need to innovate and change, to render the old obsolete and to make whatever is "new and improved" seem absolutely necessary. "Capitalism," Schumpeter wrote in a famous passage,

> is by nature a form or method of economic change and not only never is but never can be stationary. . . . The fundamental impulse that sets and keeps the capitalist engine in motion comes from the new consumer's goods, the new methods of production or transportation, the new markets, the new forms of industrial organization that capitalist enterprise creates . . . the contents of the laborer's budget, say from 1760 to 1940, did not simply grow on unchanging lines but they underwent a process of qualitative change. Similarly, the history of the productive apparatus of a typical farm, from the beginnings of the rationalization of crop rotation, plowing and fattening to the mechanized thing of today—linking up with elevators and railroads—is a history of revolutions. So is the history of the productive apparatus of the iron and steel industry from the charcoal furnace to our own type of furnace, or the history of the apparatus of power production from the overshot water wheel to the modern power plant, or the history of transportation from the mail coach to the airplane. The opening up of new markets, foreign or domestic, and the organizational development from the craft shop and factory to such concerns as U.S. Steel illustrate the same process of industrial mutation—if I may use that

---

[49]Schumpeter, *Capitalism, Socialism, and Democracy*, 83.
[50]Ibid., 82.

biological term—that incessantly revolutionizes the economic structure *from within*, incessantly destroying the old one, incessantly creating a new one. This process of Creative Destruction is the essential fact about capitalism. It is what capitalism consists in and what every capitalist concern has got to live in.[51]

In short, modern capitalism continues to be driven by a disciplined, innovative, and enterprising "spirit" that has given rise to and has subsequently been amplified by techniques and technologies of various kinds. These techniques and technologies have enabled entrepreneurs and managers to carefully adjust economic resources and the means of production to the end of profitability, and they function on the basis of the objectification of value made possible by the use of that ancient technology we call money. Linked with modern science's similar objectification of the natural world, capitalism's novel use of money has vastly increased human productivity; it has propelled the logic of enterprise and invention; and it has revolutionized life in capitalist societies and beyond. As Berger has noted, "Advanced industrial capitalism has generated, and continues to generate, the highest material standard of living for large masses of people in human history."[52]

Modern technology and the modern economy are thus crucially and inextricably linked. Indeed, the modern economy has provided the matrix within which a good deal of modern technology has developed. "In the three-cornered relations of technology, the experimental economy, and growth of material welfare," Rosenberg and Birdzell note, "the experimental economy served as a more efficient link between science and growth than any other society had achieved, and the economy was itself the source of much of its own technology."[53]

In summary, the ideology of "human progress" and/or our willingness to believe "the promise of technology" are clearly not the only factors driving modern technological development. Rather, it is also driven—and perhaps most significantly driven—by the uniquely dynamic logic of modern capitalism and its incessant demand for growth. It is often thought that the steam

---

[51]Ibid., 82-83 (emphasis in original).
[52]Peter L. Berger, *The Capitalist Revolution: Fifty Propositions About Prosperity, Equality, and Liberty* (New York: Basic Books, 1986), 43.
[53]Rosenberg and Birdzell, *How the West Grew Rich*, 33.

engine was the key modern machine,[54] and Mumford contended that it was the mechanical clock.[55] But surely money remains one of the most consequential technologies within the modern system. The evidence suggests that when you take calculating and utility-maximizing producers and consumers, give them money and the tools for keeping track of it, and place them in a relatively free market that enables them to express their preferences with money, you will end up with a kind of automatic machine that will inevitably produce both growth and change. From its inception, this economic "machine" has inspired and encouraged the development of newer, better, and more cost-effective technologies.

The Control Revolution

The fact that the economy has itself been the source of much of its own technology interprets the importance of information and communications technologies within our present context. It is no accident, in other words, that the industrial system has evolved in the direction of an "information economy." The reasons for this were explained in some detail by sociologist James Beniger. Control, Beniger began, may simply be defined as "purposive influence toward a predetermined goal."[56] Crucial to realizing predetermined goals, then, are the closely related activities of "information processing" and "reciprocal communication." Information processing entails the continuous comparison of the current state with the desired goal. Is the enterprise, its managers want to know, on track toward reasonable profitability? After comparing the current state of affairs (say, sales year-to-date) with the desired goal, if it becomes evident that course corrections are necessary, instructions for making these corrections must be communicated back to the appropriate departments. Once the corrections are made, furthermore, this new information must be relayed back to management, signaling that it is now safe to make new comparisons. And so forth. The colloquial term for this reciprocal communication is "feedback." So crucial are information processing and

---

[54]See, for example, William Rosen, *The Most Powerful Idea in the World: A Story of Steam, Industry, and Invention* (New York: Random House, 2010).

[55]Mumford, *Technics and Civilization*, 12.

[56]James R. Beniger, *The Control Revolution: Technological and Economic Origins of the Information Society* (Cambridge, MA: Harvard University Press, 1986), 7.

feedback to control that together they have become the subject of a new science called "cybernetics"—a discipline that has been principally responsible for recent advances in automation and robotics.

Beniger went on to note that the Industrial Revolution precipitated something of a crisis in control. Traditional methods of information processing and communications were simply not equal to the task of managing the increasing complexity, speed, and volume of industrial systems. It quickly became necessary to develop new strategies for controlling and managing these systems.

Relatively early on, it was realized that control could be increased, not simply by increasing a system's capacity for processing information, but also by decreasing the amount of information that actually needed to be processed. In other words, it paid to eliminate extraneous data. So it was that new techniques and methods were developed to identify and separate out relevant information, as well as to process and communicate it. One of the more important of these methods was the organizational system we commonly call bureaucracy.[57] Bureaucracy can be defined as a set of administrative procedures designed to collect and communicate relevant information to the managers of an organization, while at the same time minimizing the threats that other personalities pose to the organization.

The bureaucratization of organization was complemented with other methods and techniques similarly aimed at controlling mass production and distribution systems. Here, Beniger discussed Frederick Taylor's "scientific management," Henry Ford's assembly line, the development of automatic industrial control systems, statistical management, the science of human relations and organizational behavior, and the rise of the "supermarket."[58] Methods and techniques were also developed to attempt to manage consumption by way of market research and advertising. Taken together, all of these methods and techniques, the firms that develop and deploy them, and the people employed to operate them, comprise what we have come to call the "information economy." Contrary to fashionable opinion, Beniger concluded, microprocessing and computing technology "do not represent a new

---

[57]Ibid., 13.
[58]Ibid., 294-343.

force unleashed on an unprepared society, but merely the most recent installment in the continuing development of the Control Revolution."[59]

When Beniger wrote *The Control Revolution* in 1986, information services accounted for roughly 46 percent of GNP in the United States.[60] Putting this in historical perspective, he noted that in 1800 87.2 percent of the labor force was employed in agriculture, 1.4 percent in industry, 11.3 percent in services, and 0.2 percent in information services. By contrast, in 1980, 2.1 percent of the labor force was employed in agriculture, 22.5 percent in industry, 28.8 percent in services, and 46.6 percent in information services. Over the course of only a century and a half, the composition of the American economy was completely inverted in the direction of information and communications. Change in this direction has been particularly rapid since the 1970s, when micro-processing technologies became widely available for business use. As such technologies proliferated, and particularly with the advent of the personal computer, modern postindustrial societies have witnessed the convergence of a variety of information technologies—telecommunications, networked computing, and mass media—into what Beniger called "a single infrastructure of control."[61] As mentioned in the previous chapter, the increasingly sophisticated rendering of all the data now passing through digital networks into usable and saleable information is central to the business models of Internet giants including Amazon, Alphabet (Google), and Facebook.

Owing to the needs and logic of enterprise, then, the development of information and communications technologies appears to have been all but inevitable. Equally inevitable, Beniger suggested, was the disappearance of traditional cultural mores and customs that had accompanied older control strategies. Along this line, Beniger contended that "family partnerships, the Southern 'gentleman's code,' [and] the community relations that bound farmer and storekeeper [survived] only so long as technologies of information processing, transportation, and communication remained undeveloped."[62]

---

[59]Ibid., 435.
[60]Ibid., 22.
[61]Ibid., 25.
[62]Ibid., 166-67; see also Peter Berger, "On the Obsolescence of the Concept of Honor," in *Revisions: Changing Perspectives in Moral Philosophy*, ed. Stanley Hauerwas and Alasdair MacIntyre (Notre

As effective strategies for control in these areas were devised and deployed, more personal means of control atrophied quickly. Traditional mores and customs may have held certain cultural advantages, Beniger observed, but they were simply not efficient in the modern sense.

Finally, and in an attempt to interpret the seemingly inexorable movement toward increasingly sophisticated information and communications technologies, Beniger suggested that the development simply mirrors the survival strategy of living systems in the face of entropy. "In order to oppose entropy," he noted,

> and put off for a time the inevitable heat death, every living system must maintain its organization by processing matter and energy. Information processing and programmed decision are the means by which such material processing is controlled in living systems, from macromolecules of DNA to the global economy. The Industrial Revolution sharply increased the volume and speed of energy conversion and material processing and thereby precipitated the various technological responses we call the Control Revolution. Even if industrialization had been more gradual, however, the ultimate result would have been much the same.[63]

The so-called digital revolution, with all of its disruptive advances in information processing, communications, cybernetics, automation, and robotics, owes its existence finally and very largely to the requirements of the modern economy. These same requirements suggest that innovation will likely continue to occur in the direction of fully automated control technologies. In any number of important respects, modern business and industry are rapidly outgrowing the need for human administration and interpersonal communication.

Other Connections

That modern technology has by and large developed within the matrix of capitalism explains its predominantly bourgeois flavor. Although the Promethean possibilities of modern technology occasionally surface in research laboratories and military applications,[64] modern technologies are more

---

Dame, IN: University of Notre Dame Press, 1983), 172-81.

[63]Ibid., 58-59.

[64]Of course, even the development of military technology reflects the logic and demands of the market economy, for the high costs of developing new weapons technologies are very often

commonly directed toward the ordinary desires of average consumers for such things as creature comforts, convenience, and safety. Democratic capitalism, it seems, ensures that smartphones, consumer-electronic gadgetry, and the production of action-adventure films outweigh more audacious ambitions on the list of contemporary technological priorities. While the transcendent aspirations of so-called transhumanists may well be taken up in films, and although transhumanism does periodically attract the attention of the mass media, it is the concerns of ordinary people that actually attract investment. The downside to this, of course, is that trying to steer modern technological development away from the expressed preferences of ordinary consumers for safety, comfort, and convenience in light of, say, looming ecological problems, will not be easy.

The fact that a good deal of modern technology has developed within the context of the market economy also helps to interpret the increasingly dizzying pace of technological change, as well as the restlessness and insatiability that characterizes contemporary culture. We will have more to say about technology's restructuring of time in our fifth chapter. Here, we simply note that automatic machine technology is today being resourced, developed, and deployed within an economic system in which "time is money." The venture capitalists who are even now searching for the next "killer app" and/or the next "game-changing" technology could not for the most part care less what the "app" will actually do or if "the game" is changed for the better. Their concern is simply to see return on investment, which can then be reinvested in the next "next big thing." The process of "creative destruction" and the market system's apparently limitless appetite for novelty ensures that investing in whatever is next is almost always a good bet. And the more quickly this bet can be placed, the better.

The capitalist context also helps to explain why modern technological development often seems to display no final purpose beyond that of simply delivering "more." In this, it simply reflects the logic of money—for, of course, money can be indefinitely increased. In attempting to interpret the enormous

---

allayed by selling slightly downgraded versions of these technologies to whomever is willing to pay for them. Military planners have learned a basic but very valuable lesson from the civil economy: the more units sold, the lower the production costs per unit.

upsurge in industrial output after 1750, Mumford suggested that it had more to do with the increasing influence of money upon society than with technological developments per se. Things in nature, he observed, inevitably encounter limits. They grow and develop only insofar as they are able to fit within a given environment, and only insofar as their growth and development are consistent with the physical requirements of life. "But when human functions are converted into abstract, uniform units," Mumford noted, "ultimately units of energy or money, there are no limits to the amount of power that can be seized, converted, and stored. The peculiarity of money is that it knows no biological limits or ecological restrictions."[65] When growth is measured almost exclusively in monetary terms, as it is within the capitalist system, it is for all intents and purposes limitless. Thus, whereas it is often assumed that modern technological development drives economic growth, it is perhaps truer to say that our technologies, like everything else in the context of the capitalist economy, are being swept along by the system's incessant need to grow.

That so much of our technology has been developed within the matrix of the market economy also helps to explain the increasing plausibility a kind of soft nihilism within contemporary technological culture. More and more of our contemporaries, it seems, appear to believe that reality itself is not only malleable but that it is an ultimately artificial and arbitrary human construct. There are very good history-of-ideas explanations for the emergence of this kind of belief, and many attribute it to the secularization of modern societies. Yet the plausibility of nihilism today is surely also underwritten by the fact that we are almost completely surrounded by human-made objects and are almost constantly immersed in artificial and increasingly "virtual" environments.

Nihilism's plausibility today is surely also a reflection of our internalization of the money economy.[66] Simmel observed this phenomenon in a celebrated essay entitled "The Metropolis and Mental Life" (ca. 1900):

The modern mind has become more and more calculating. The calculative exactness of practical life which the money economy has brought about corresponds to the ideal of natural science: to transform the world into an

---

[65]Mumford, *Myth of the Machine*, 165.
[66]Simmel, "Metropolis and Mental Life," 414.

arithmetic problem, to fix every part of the world by mathematical formulas. Only [the] money economy has filled the days of so many people with weighing, calculating, with numerical determinations, with a reduction of qualitative values to quantitative ones.[67]

As we have become more and more calculating, Simmel went on to note, so we have also become cynical and blasé; money has leveled our experience of the world by making everything conveniently comparable in terms of simple arithmetical calculations. We have come to experience the world as a place devoid of qualities, an increasingly dreary place in which, as Simmel put it, "all things lie on the same level and differ from one another only in the size of the area which they cover."[68] Under such circumstances, money— which is supposed to be simply a means to an end, one tool among many— slides out in front of all of our other purposes and displaces them all in a kind of "teleological dislocation."[69] Money becomes the only purpose. Søren Kierkegaard already captured this odd modern development with characteristic irony at the beginning of the nineteenth century:

> In the end, therefore, money will be the one thing people will desire, which is moreover only representative, an abstraction. Nowadays a young man hardly envies anyone his gifts, his art, the love of a beautiful girl, or his fame; he only envies him his money. Give me money, he will say, and I am saved. . . . He would die with nothing to reproach himself with, and under the impression that if only he had had the money he might really have lived and might even have achieved something great.[70]

Thankfully, we have not gotten to the point where money has become the only purpose for most people. But we do appear to be moving in the direction of deeming our representations of reality, and our abstractions from it, more important than reality itself. Just as the pervasive use of money has encouraged us to transform our world into a series of arithmetic problems to which "more" is always the solution, so it seems that our remarkable technical skills have convinced us that "reality" is something that we can redesign.

---

[67]Ibid., 412.

[68]Ibid., 414.

[69]Simmel, *Philosophy of Money*, 526.

[70]Søren Kierkegaard, *The Present Age and Of the Difference Between a Genius and an Apostle*, trans. Alexander Dru (New York: Harper & Row, 1962), 40-41.

The close connection between modern technology and the modern economy also suggests, as indicated above, that the technological society may be difficult to reform. For not only does the system function almost automatically to generate profits and innovation, but all of us depend upon it to continue generating both. With respect to profits, the burden of proof today falls upon anyone who would suggest that there ought to be some basis other than that of rational self-interest on which to make business decisions. After all, individuals and firms must pursue their own self-interest in a rational fashion for the market system to function efficiently and for the economy to continue to grow. The economy must continue to grow, furthermore, if it is to continue to attract the capital of rational and self-interested investors and if it is to support the material aspirations of all who participate and have a stake in the system. An equally heavy burden of proof falls upon anyone who is perceived to stand in the way of technological innovation. Technological innovation has long been—and remains—one of the keys to economic growth and prosperity.

Excursus: "La Technique" or "Rationalization"

The apparently indivisible combination of science, technology, and the economy persuaded noted Christian philosopher Jacques Ellul to insist that the technological society is actually governed by a single overarching rule, a kind of governing method that he sought to capture in the term "*La Technique*." Ellul defined "*La Technique*" as "the totality of methods rationally arrived at and having absolute efficiency (for a given stage of development) in every field of human activity."[71] While concise, Ellul's definition is famously opaque. It raises a number of questions: "Rational" in what sense? In what way "efficient"? In "every" field of human activity? What is meant by "a given stage of development"?[72]

Such questions are helpfully seen in light of Max Weber's somewhat earlier attempt to capture the essence of modernity under the heading of

---

[71]Jacques Ellul, *The Technological Society*, trans. John Wilkinson (New York: Vintage, 1964), xxv.
[72]I developed this material in more detail and at greater length a chapter entitled "The Intrinsic Secularity of Modern Economic Life," in *The Way of the (Modern) World: Or, Why It's Tempting to Live as if God Doesn't Exist* (Grand Rapids: Eerdmans, 1998), 131-79.

"Rationalization." Weber's understanding captures the essence of modern development even as it suggests the kinds of questions we may need to ask if we want to comprehend technique from beyond its own dynamism. Weber insisted that modernity is animated by a process in which society and culture are progressively surrendered to rational methods and techniques. Sociologist Thomas Luckmann summarized this position:

> In Weber's view the outstanding characteristic of modern society is its "rationality." . . . In modern society he discerned the prevalence of a highly systematic, anonymous and calculable form of law, he found an economy guided by its own principles of calculability and means/ends rationality, he observed a trend to an anonymous, calculable and bureaucratic system of political administration and, last but not least, he of course noted the social significance of an objective science that made nature technically calculable. . . . The "rationality" of modern society was perceived by Weber as the result of a unique line of historical development which—as soon as it became welded into the structure of society—became divorced from the conditions of its origin and either overwhelmed all other lines of historical development or came to serve as a model for them.[73]

The "conditions of its origin" to which he refers were Judeo-Christian, and the "other lines of historical development" that have by now been overwhelmed by the process of rationalization were also, for the most part, Judeo-Christian. Thus, whereas Christianity had provided a kind of overarching interpretive canopy over Western European society and culture for centuries, with the onset of modernity European social and cultural life began to be reinterpreted along purely secular lines and surrendered to "rational" methods developed on the basis of "scientific" understanding. This process didn't happen overnight, and in a number of important respects, it is still happening. Yet most of the central institutions in modern Western societies no longer function, either theoretically or practically, with explicitly religious ends in view. Instead, they have been envisioned and designed to guarantee penultimate or secular benefits. Modernity's central institutions are now geared almost entirely toward delivering things like prosperity,

---

[73]Thomas Luckmann, "Theories of Religion and Social Change," *The Annual Review of the Social Science of Religion* 4 (1980): 14.

safety, and convenience. All of this, Weber believed, was a reflection of the modern process of rationalization.

"Rational" conduct, according to Weber (and in common usage generally), simply entails choosing means suitable to realizing a desired end. Along this line, he distinguished rational action from conduct prescribed unthinkingly by tradition and/or custom, as well as from purely emotive behavior. He also understood the term "rationality" to be of rhetorical significance in the sense that it can be attached to actions after the fact in order to justify them. Rationality is an expression of human desire as well a purely intellectual achievement.

This brings up the question of what is worth seeking by way of rational action. After all, few things are rational in and of themselves. Rather, they are seen to be so, Weber notes, "from a particular point of view. For the unbeliever, every religious way of life is irrational, for the hedonist, every ascetic standard."[74] Yet for the Christian, foregoing this-worldly pleasures for the sake of one's soul is supremely rational. Like beauty, it seems, rationality lies in the eye of the beholder; it all depends on where one is trying to go and what one is trying to accomplish.

Weber attributed the perspectival character of rationality in part to the fact that there are actually several different types of it, each type oriented toward a different end. The different types he identified (though these are not the words he used) were theoretical, formal, substantive, and practical.[75] Theoretical rationality seeks to make sense of the world by means of theoretical abstraction. Modern science provides a good example of this, for within modern scientific understanding it is just as important to be able to explain why things happen as it is to observe that they happen. Formal, or official rationality, according to Weber, is largely (though not entirely) unique to the modern situation and aims at the preservation of formally established organizations by means of carefully defined roles, rules, and procedures. Bureaucracy is the fruit of formal rationalization. Substantive rationality entails the active shaping of life, both practically and theoretically, such that it may be

---

[74]Weber, *Protestant Ethic*, 194n9.
[75]See Stephen Kalberg, "Max Weber's Types of Rationality: Cornerstones for Analysis of Rationalization Processes in History," *American Journal of Sociology* 85 (1980): 1145-79.

brought into accord with religious and/or ethical values. Last, practical rationality simply entails trying to realize practical, typically material, ends.

Each of the different types of rationality has historically given rise to different episodes of rationalization, episodes that have for the most part occurred independently of one another. Weber discussed a number of examples, such as the formal rationalization of law in ancient Rome and, following on from this, the formal (bureaucratic) rationalization of canon law within the Roman Catholic Church. Weber also noted that practical rationality, geared as it always is toward the pursuit of merely material ends like wealth, has almost always been subjected to theoretical, formal, and substantive discipline. Economic action in the medieval period, for example, and as noted earlier, was closely regulated by formal church law, which was itself informed by substantive theological reasoning about such things as just prices and the prohibition of usury. What makes modernity unique, Weber believed, is that, owing to a unique line of historical development, all of its central institutions have been rationalized simultaneously and synergistically such as to be aimed at the realization of mundane, practical, and material purposes. A relatively narrow and somewhat blinkered secular focus has thus come to characterize modern societies, making what we have called "the promise of technology" to liberate us from want and drudgery appear both alluring and compulsory. This is the crucial piece of the modern puzzle that Ellul's sweeping definition of "*La Technique*" tends to obscure: it is not simply that modern methods and systems strive after efficiency but rather that they strive after the efficient realization of ends that are almost exclusively practical, material, and this-worldly.

Thus, while the process of rationalization always strives to master reality in some meaningful way, in the modern context the mastery most commonly sought is practical and material. We seek to achieve this mastery, as we have seen, by way of things like careful empirical investigation, information processing, and reciprocal communication. There is no doubt that rationalization makes our understanding of things more powerful. Yet it also tends to specialize and narrow our understanding.

The assumption that reality can be mastered also contains the implicit metaphysical assumption that no mysterious and/or incalculable forces bear upon our spheres of interest and activity. This assumption necessarily requires the implicit—and very often explicit—rejection of tradition, custom, religion, and indeed, personality as providing an adequate basis for rational decision making. After all, if planners were to admit of such things as grace, mystery, miracle, personality, and freedom, forecasting and prediction would become difficult if not impossible.

This is why the modern world has become, to use Weber's famous term, so thoroughly "disenchanted." As Weber noted,

> The increasing intellectualization and rationalization do not, therefore, indicate an increased and general knowledge of the conditions under which one lives. . . . It means something else, namely, the knowledge or belief that if one but wished one could learn it at any time. Hence it means that principally there are no mysterious incalculable forces that come into play, but rather that one can, in principle, master all things by calculation. This means that the world is disenchanted. One need no longer have recourse to magical means in order to master or implore the spirits, as did the savage, for whom such mysterious powers existed. Technical means and calculations perform the service. This above all is what intellectualization [and rationalization] means.[76]

In sum, Weber believed that rationalization lies at the heart of modernity, and that the process of modernization consists in the rationalization of more and more spheres of social life, a process that tends to cascade as the rationalization of one sphere spills over into others. As a result, virtually all of modern life, from managing the national economy to sending people into space, to curing diseases, to raising children, is surrendered to rational methods and techniques. The whole rationalizing process seems to be aimed at the predictable delivery of secular benefits such as health, prosperity, safety, security, comfort, and convenience.

The United Nations' "Millennium Development Goals" (set in 2000)—to eradicate extreme poverty and hunger, to achieve universal primary education, to promote gender equality and empower women, to reduce child mortality, to improve maternal health, to combat HIV/AIDS, malaria, and other diseases, to ensure environmental sustainability, and to develop a

---

[76]Max Weber, "Science as a Vocation," in *From Max Weber*, ed. H. H. Gerth and C. Wright Mills (New York: Oxford University Press, 1946), 139.

global partnership for development[77]—well describe the secular scope of distinctively modern aspirations, and most are rather easily recast as technical problems for which technological solutions need now to be developed. It is hardly surprising, then, that technology figures so centrally within the modern context, and it is no wonder that Ellul believed that ours is a "technological society" that is all but defined by the totality of rational methods governing practically every field of human activity. For ours is a society in which taking control of our secular circumstances by means of rational-technical means, methods, procedures, and techniques has become supremely important.

Concluding Remarks

The momentum and inertia of modern technological development derive in large part from its close integration, not merely with modern scientific progress, but with powerful economic forces and interests. While technology and economics have different histories prior to the Industrial Revolution, the two have been inextricably intertwined since the middle of the eighteenth century. Just where the one leaves off and the other begins is often difficult to determine. This is why the outlook of the typical technological worker described in the last chapter—characterized by "problem-solving inventiveness" and reliance upon "expertise"—can just as accurately be said to describe the outlook of the typical businessman. It is also why certain problematic aspects of modern life that are commonly blamed on technology—its increasingly frenetic speed, for example, or the "objectification" of people—should also to be seen to reflect economic forces and the insidious ramifications of the money economy.

The close connection between modern technology and modern capitalism helps to explain why so many recent technological developments have occurred in the areas of information processing and communications —and also why, aside from military applications, technological developments have tended to converge on the perceived needs and demands of ordinary consumers, as well as on the manipulation of consumer desire.

---

[77]United Nations, www.unmillenniumproject.org/goals/.

The close technology-capitalism connection explains the pace of technological change as well as the restlessness and insatiability that seem to characterize the technological society. It would also seem to interpret modern society's secularity. Indeed, the close connection between modern technology and the modern economy sheds light on the plausibility of what Charles Taylor has called "exclusive humanism," that peculiarly modern ethos that entertains no final human purposes beyond those of material flourishing in the here and now.[78]

A somewhat less charitable label for exclusive humanism is nihilism, that decidedly modern conviction that there simply is no truth or meaning beyond that which we construct and/or engineer for ourselves. Such a conviction could perhaps only have gained widespread plausibility within an environment that had become so saturated with human artifice that the heavens—now obscured by smog and fluorescent lighting—were no longer able declare the glory of God (Ps 19:1).

Unsurprisingly, the nihilism subtly encouraged by a technological society cannot offer much resistance to further technological development. Indeed, as cultural historian Leo Marx has noted, there is actually a close affinity between the postmodern conception of power and the functioning of large technological systems.[79] Both are dynamic, fluid, and diffuse, with no single, fixed, or even discernable locus, and neither can be resisted in the name of normativity or, as Marx puts it, for the sake of "a larger political vision of human possibilities."[80] Even to attempt to do so, so the postmodern argument runs, would be dangerously "protototalitarian."[81] In its diminished sense of human agency, Marx laments, the postmodern mood acquiesces in the face of modern technological development and often amounts to a kind of "ambivalent tribute to the determinative power of technology."[82]

Even where moral and ethical questions are earnestly raised by contemporary critics of technology, they are most often answered instrumentally, usually in terms of "values" that we are entreated to choose in order to solve

---

[78]Charles Taylor, *A Secular Age* (Cambridge, MA: Harvard University Press, 2007), 245.
[79]Leo Marx, "The Idea of 'Technology' and Postmodern Pessimism," in Leo Marx, *Does Technology Drive History? The Dilemma of Technological Determinism* (Cambridge, MA: MIT Press, 1994), 237-57.
[80]Ibid., 257.
[81]Ibid. Marx writes that he is thinking primarily of François Lyotard and Michel Foucault here.
[82]Ibid.

the problems that our technologies have inadvertently occasioned.[83] Such protests offer no real vantage point from which to judge modern technological development from beyond its own internal logic. They are, in any event, as nothing compared to the will-to-power that animates the modern technological/economic system.

Yet having considered some of the effects that our technologies appear to be having on us, and having now examined modern technology's close ties to economic forces and interests, it remains for us to try to determine just how we have gotten to this place of believing that it is up to us to assign "values" to the stuff of the world. How, put differently, have we arrived at the point where we can imagine no other way of being-in-the-world beyond that of relentless engineering? The next chapter will delve into this question by way of the history of ideas. This ideational history will hopefully prepare us to apprehend and appreciate the tremendous implications of Christ's incarnation for the technological society.

---

[83]See, for example, Bill Joy, "Why the Future Doesn't Need Us," *Wired*, April 2000, www.wired
.com/2000/04/joy-2/.

CHAPTER THREE

THE TECHNOLOGICAL WORLDVIEW

*The question underlying our bewilderment about being human [today]
... may well be how we came to convince ourselves of the putative truth
of the deeply counterintuitive and counterhistorical notion of the human
as a stranger contingently thrown into a meaningless, mechanical world.*

ERAZIM KOHAK, *THE EMBERS AND THE STARS*

FROM THE BEGINNING, machine technology was developed to function automatically. That is to say, it has been designed and deployed to function independently of unauthorized human interference and unimpeded (as much as possible) by human frailties, inconsistencies, and irrationalities. Automatism, we said, is one of the keys to machine technology's utility, in large part due to the economic need to control costs. Modern technological development has as a result been moving away from ordinary embodied human existence for some time. Indeed, the image of the strangely alien, independent, and often menacing machine has, since the beginning of the nineteenth century, haunted the modern imagination and all but defines the modern literary genre of science fiction. It has become a staple of dystopian fiction to envisage the dire implications that the development of fully automatic—and fully independent—machinery could have for human life. Still, as automatic as our machines have become, real people have continued to design, operate, repair, and maintain the machinery of modern life. Even our so-called automobiles, to cite one example, still require human drivers.

If certain futurists are correct, all this is about to change. Newly augmented by computerized and networked command and control systems, modern machine technology is now poised at long last to become genuinely automatic. The orders we place online, for example, are increasingly for products that are made, picked, packaged, and may soon—if Amazon.com builds its fleet of drones—be delivered to our doorstep entirely by machines. Everyday services, from banking to buying groceries, to making travel arrangements, to routine visits to the doctor's office, soon will be managed largely by machinery. Reliable "driverless" cars are even now being tested on public roadways.

One need not resort to images of Skynet or *The Matrix* to suspect that that the influx of algorithms and fully automatic machinery into modern life may not after all be the best news for real people. While perhaps further reducing costs and making products and services easier to consume, the automation of industry is going to eliminate the need for human laborers in large swaths of the modern economy. Of course, we can hope that those who are thus displaced will find interesting and perhaps more satisfying work, but this is beginning to appear increasingly unlikely. What, one wonders, would anyone be willing to pay them to do? This is one of the reasons economists and even technology insiders are beginning to voice grave concerns about the directions that fully automatic machine technology appears to be taking us.[1]

What is surprising is that most of the rest of us don't appear to be too concerned. Yes, it's frightening to imagine a world in which human beings have been enslaved by machines, but that's simply the stuff of Hollywood. Surely, we—or whoever is in charge—will not let matters come to that. And in the meantime, the impact that automatic machine technology is already having upon us—the self-anonymization, the myopia of specialization, the confusion of means and ends, the loss of skills, the psychological fallout—is cause for concern, but not yet for alarm. These problems, we tell ourselves, will surely be redressed in due course, most likely with newer and better technologies.

---

[1] See Erik Brynjolfsson and Andrew McAfee, *The Second Machine Age: Work, Progress, and Prosperity in a Time of Brilliant Technologies* (New York: W. W. Norton, 2014); also Jaron Lanier, *Who Owns the Future?* (New York: Simon & Schuster, 2013); see also Bill Joy, "Why the Future Doesn't Need Us," *Wired*, April 2000, www.wired.com/2000/04/joy-2/.

Just why we are so accepting of automatic machinery, and why we are not more concerned that modern technological development is trending away from ordinary embodied human existence—which is to say, away from *us* (!)—is the subject of this third chapter. Why, putting the question slightly differently, has there not been more of an outcry that, as Neil Postman lamented a number of years ago, technology has become the sole substance of modern culture?[2] Or, to paraphrase Martin Heidegger's celebrated observation (to which we will return), how is it that we have allowed technology to become, in effect, the metaphysic of the modern age? This chapter's thesis is that our naive lack of concern stems from a particular view of the world, a view that sees nature—including human nature—as a vast and elaborate mechanism, one that differs from automatic machine technology in scale and intricacy but not in kind. Ordinary embodied human existence from within this worldview is, again, most often construed not as something to be nurtured and enhanced, perhaps by technology, but rather as a series of limitations that remain to be overcome with more and better technology. From within this technological worldview, human embodiment is simply not a particularly high priority. This, I want to suggest, betrays serious confusion about the nature of the created order as well as confusion about the human place and task within the created order.

The genealogy of the technological worldview can be approached from a number of different angles. As we have seen, simply using modern technology tends to foster an outlook that construes life as a series of discrete problems that are to be solved by the application of discrete techniques. The technological workplace, as we have also discussed, both requires and rewards a kind of problem-solving inventiveness that assumes that work and life can be understood in a largely mechanical fashion.

Money's pervasive leveling of the world, furthermore, and its relentless reduction of quality to quantity, is both consonant with and has actually accelerated the modern drive toward more and more automatic machinery. If we find that real people are increasingly being replaced by machines in the workplace, this is simply because machines make more sense in terms of the

---

[2]Neil Postman, *Technopoly: The Surrender of Culture to Technology* (New York: Vintage, 1993).

rational cost-profit calculus. As we saw in our last chapter, powerful economic forces and interests drive the "creative-destructive" process of technological change in the direction of automation.

Of course, it may also be the case that, in failing to realize that we are effectively being replaced by the very technologies that we often embrace in the name of human progress, we have fallen into what author Ronald Wright called, in *A Short History of Progress* (2005),[3] a "progress trap." This, Wright contended, is the perplexing predicament that a society finds itself in when, although the need for systemic change might well be obvious, no one is willing to risk fundamentally changing a status quo that they have come to depend upon. Wright's book grew out of a series of lectures commissioned by the CBC (Radio-Canada) in 2004, in which he drew attention to a number of complex and apparently successful civilizations that collapsed in the face of ecological and social problems that they must surely have seen looming. Using the fallen civilizations of Easter Island, Sumer, Rome, and the Mayan peninsula, as well as examples from the Stone Age, Wright suggests that societies can become trapped by their own development as they encounter unanticipated problems that they do not possess the resources and/or political determination to solve. Seeking to avoid losses in quality of life, social stability, status, and relative standing any number of once-thriving human societies have apparently failed to redress the social and environmental problems that their own development—their "progress"—created. This has happened in spite of the fact that some of these problems—such as the elimination of all of the trees from Easter Island—must have been painfully obvious to all concerned. Failing to redress these problems, these societies inevitably collapsed. As Wright observed,

> Many of the great ruins that grace the deserts and jungles of the earth are monuments to progress traps, the headstones of civilizations which fell victim to their own success. In the fates of such societies—once mighty, complex, and brilliant—lie the most instructive lessons. . . . they are fallen airliners whose black boxes can tell us what went wrong.[4]

---

[3]Ronald Wright, *A Short History of Progress* (New York: Carroll & Graf, 2004).
[4]Ibid., 8.

While the dire repercussions of progress traps have historically been regional, what, Wright wonders, if our increasingly global civilization were to fall into such a trap? What if it already has?

While it is undoubtedly true that our aversion to short-term losses in status, stability, and/or quality of life has left us somewhat insensitive to the drift of modern machine development away from ordinary, embodied, human reality, our reluctance to critically scrutinize modern technology would seem to run deeper than the fear of loss and/or political indecision. After all, the general consensus today is that increasingly automatic machine technology is the solution to many of our problems, not the problem itself.[5] The key resource we appear to lack in confronting modern technological development is not political courage. Rather, it is our inability to imagine any other way of being-in-the-world beyond that of problem-solving inventiveness. As Horowitz and Grant observed, a technological society is one in which people

> think of the world around them as mere indifferent stuff which they are absolutely free to control any way they want through technology. . . . [I]t is a whole way of looking at the world, the basic way western men experience their own existence in the world. Out of it come large organizations, bureaucracy, machines, and the belief that all problems can be solved scientifically, in an immediate quantifiable way. The technological society is one in which men are bent on dominating and controlling human and non-human nature.[6]

We will unpack these observations below. Here at the outset of our discussion, simply note that there are a number of basic beliefs implicit in this way of looking at the world. The first is that the world is best understood as a vast collection of component parts or elements that are interconnected in a more-or-less mechanical fashion. Modern science reveals what these elements are and the laws by which they interact. Having determined how the world works by means of modern science, then, we are free to reconfigure or "engineer" it for the sake of realizing purposes that we are entirely free to choose. From this perspective, nature is understood as a kind of vast machine, and our bodies are simply part of the machinery, available for

---

[5]See, for example, John Asafu-Adjaye et al., "An Ecomodernist Manifesto," on the website www .ecomodernism.org/.

[6]Gad Horowitz and George Grant, "A Conversation on Technology and Man," *Journal of Canadian Studies* 4 (August 1969): 3.

enhancement, improvement, and/or augmentation. That such a worldview should foster instrumentalism, functionalism, and the engineering mentality should come as no great surprise, for it all but requires the objectification of both nature and people. Continuing along this line, it is not terribly surprising that various nihilisms have surfaced within modern technological culture, for the modern—and, for that matter, postmodern—way of looking at the world simply assumes that the world has no intrinsic worth apart from the imposition of our will (often hidden behind the language of "values") upon it. In short, it is no real wonder that modern men and women are not more alarmed by the disembodied drift of modern machine development. We have become accustomed to thinking about the world—and, indeed, about ourselves—on the analogy of machinery.

It has been suggested in this connection that our modern technological worldview is simply a modern-day variation on that ancient theme of gnosticism, that hoary amalgamation of religious convictions and practices whose adherents shunned the lower regions of mere physicality and materiality and sought release into an upper, purely spiritual realm by means of initiation into special religious insight and knowledge (*gnosis*). Ancient gnostics were known for disparaging their bodies, either by way of masochistic ascetic practices or in licentiousness and debauchery. Gnosticism parading as true Christianity was one of the first counterfeits the early church had to confront, and it did so by way of robust theologies of creation and incarnation. The early church recognized that beliefs and practices that devalued ordinary embodied human existence must be seen to be at odds with God's redemption of created order.

Now, although modern-day technological "gnostics" do not typically contend that matter per se is evil—though some have argued that organic matter is unnecessarily fragile—they do tend to chafe at the traditional religious conviction that created nature ought to shape and delimit human aspirations. On the contrary, as it is sometimes asserted today, nature—including the human body—is something to be mastered by the human spirit and altered if necessary for the sake of purposes that human beings have willed. Gnostic themes are for the most part only implicit in the modern technological worldview, but they have surfaced explicitly and with increasing frequency at the forefront of technological research and development. Beliefs akin to gnosticism are very much at the heart of the so-called cybernetic revolution, the belief that it is possible to reduce all of material reality to slivers of underlying information that can be digitized and thereafter infinitely manipulated and reconfigured. Given the cybernetic view of the world, ethicist Brent Waters observes in an interesting study entitled *From Human to Posthuman* that the only thing standing in the way of the radical transformation of nature—including human nature—is inadequate technology, a problem that can be remedied, at least in principle, by further research and development. "There is no real boundary separating nature and artifice," Waters comments, "only patterns or lines of information that can be erased and redrawn; no real limit that cannot be eventually overcome."[7]

Such a view of the world may once have been the stuff of science fiction, but it is expressed with increasing seriousness today. Just why this is, and why modern and postmodern thought has tended to erase the boundary between given nature and human artifice, is something we must try to understand. For in erasing this boundary, modern and postmodern thought has lost sight of the intrinsic value of ordinary embodied existence as well as of the distinctively human vocation within the created order.

The Mechanical World Picture

As we noted in the last chapter, Lewis Mumford began his magisterial history of technological development, *Technics and Civilization*, with the contention that a kind of mechanical world picture antedated the development of recognizably modern machinery by several centuries.[8] What had formed the foundation for the spectacular technological achievements of the last two centuries, Mumford insisted, were not simply scientific discoveries and/or novel solutions to technical problems but rather a change of *mind*, a new way of looking at the world and a new estimation of human purposes within it.

---

[7]Brent Waters, *From Human to Posthuman: Christian Theology and Technology in a Postmodern World*, ed. Roger Trigg and J. Wentzel van Huyssteen, Ashgate Science and Religion Series (Aldershot, UK: Ashgate, 2006), x.

[8]Lewis Mumford, *Technics and Civilization* (London: Routledge & Sons, 1934).

Mumford went on to elucidate a number of examples of this new way of seeing the world: the regularizing of time, space, distance, and movement, which, together and by means of various mathematical operations, enabled early scientists and engineers to objectify, explain, and predict all manner of natural processes. This objectification of the world of nature was powerfully reinforced by the quantitative logic of money.[9] Here again, we note that an experimental economy appears to have served as the crucial matrix within which science, technology, and the growth of material welfare were effectively linked. In his later work, Mumford believed that one could actually pinpoint when these various developments coalesced to form the new worldview. "If there is any one point," he wrote,

> at which one may say the modern world picture was first conceived as the expression of a new religion and the basis of a new power system, it was in the fifth decade of the sixteenth century. For not merely was Nicolaus Copernicus' 'De Revolutionibus Orbium Coelestium' published, but likewise Vesalius' treatise on anatomy, 'De Humani Corporis Fabrica' (also 1543), Jerome Cardan's algebra, 'The Great Art' (1545), and Fracastoro's enunciation of the germ theory of disease, 'De Contagione et Contagiosis Morbis' (1546). Scientifically speaking, that was the decade of decades: unrivalled until our own century.[10]

Mumford also observed that the mechanical world picture had been introduced into North American culture together with a kind of Romantic pastoralist reverence for nature that for a time held the former in check. Yet the power over nature conferred by new modern machinery as well as the profits such power augured eventually proved irresistible. North American culture quickly came to be dominated by the logic of commercial and technological development. So much so, in fact, that by the end of the nineteenth century, "a large portion of the human race would virtually forget that there had ever existed any other kind of environment, or any alternative mode of life."[11]

Mumford recognized that the modern technological milieu, an environment increasingly dominated by machines and machine systems, would

---

[9]Ibid., 25.
[10]Lewis Mumford, *The Myth of the Machine: The Pentagon of Power* (New York: Harcourt Brace Jovanovich, 1964), 29.
[11]Ibid., 24.

have been largely unimaginable to the early-modern thinkers most responsible for shaping the mechanical world picture. Galileo, for example, surely
could not have suspected that the mechanical worldview would result in "an
environment like our present one: fit only for machines to live in."[12] Still—at
least we can say this in retrospect—it is hardly surprising that a kind of
mechanical outlook might have given rise to the development of machinery,
and eventually to a kind of mechanical social order. In what follows, then, I
want to highlight several crucial episodes in the evolution of the mechanical
world picture. If it is true that people became mechanical in outlook before
they developed the machinery to express their new "bent and interest," then
trying to understand the ideational history of the mechanical outlook represents an important step toward evaluating and possibly redirecting
modern technological development toward less mechanical—and hopefully
more *humane*—ends.

Cartesianism

French polymath René Descartes (1596–1650) is often credited with formu-
lating the distinctively modern view of nature as a machine. Critical of prevailing philosophical and/or theological descriptions and explanations of
natural phenomena, Descartes determined to ground his understanding of
nature upon a new foundation, and he encouraged his seventeenth-century
readers to envision nature as a vast and intricate mechanism. Descartes insisted that the traditional practice of seeking to identify, and then to reason
from natural "essences," an endeavor dating back to Aristotle, as well as the
Aristotelian habit of seeking to explain natural events in terms of "final" or
ultimate causes—two intellectual practices that had long animated the discipline of natural theology—were in practice anthropomorphic and "occult."
Both imposed debatable theological meanings upon natural processes and
added nothing useful to our knowledge of the world. In fact, and in spite of
nature's apparent variety, Descartes insisted that the natural world was comprised of the same basic material elements, bits of natural "stuff" that occupy
space and that collide with one another to produce variety and pluriformity.

---

[12]Ibid., 57.

What appear to be unique qualities—colors, sounds, tastes, textures, and so on—are simply the product of these collisions and are therefore entirely explicable in terms of mass, speed, and other values that are mathematically representable. Once described mathematically, natural processes can be predicted—and potentially, at least in principle, controlled. The point of Descartes's "new philosophy," then, was to render us "the masters and possessors of nature."[13]

Curiously, those who thus master and possess nature because they understand its inner workings are not themselves part of the natural mechanism, but are conceived to stand outside the world of natural objects, observing them from above as "thinking substances" or knowing "subjects." The relations of human observers to the world of nature are thus understood on the analogy of the relation that God is thought to have to his creation. As William Poteat has observed,

> [The Cartesian outlook] is comprised of a coherent system of mutually implicative images, metaphors, and analogies that represent man's relation to nature, to his own body, to the world of material objects, to time and history, to his acts of reflection, to his decisions, to his intellect, *even to his own ego*; and these relations are analogous to the relation that God is conceived to have to the world that he has made out of nothing. Man is here depicted, in other words, as essentially disembrangled from, because transcendent over, and thus autonomous in relation to all of these.[14]

Hence the distinctively modern notion of "worldview," which suggests the perspective of one who is somehow able to stand entirely outside of the world and to look back at it. Hence also the well-known Cartesian dualisms that separate knowers from what is known, subjects from objects, the self from others, and so forth.

As indicated in Poteat's comments above and as Charles Taylor explains in *Sources of the Self: The Making of the Modern Identity*, the Cartesian project entailed a kind of "self-objectification," which is to say, the viewing

---

[13]Descartes, cited in John Passmore, *Man's Responsibility for Nature: Ecological Problems and Western Traditions* (New York: Charles Scribner's Sons, 1974), 20.

[14]William H. Poteat, *Polanyian Meditations: In Search of a Post-Critical Logic* (Durham, NC: Duke University Press, 1985), 252-53 (emphasis in original).

of oneself as if from a third-person perspective. Our bodies, after all, are parts of nature. They must also, therefore, be mastered and possessed.[15] In this connection, Taylor notes that the classical goal of self-control, in which one sought to master the passions for the sake of realizing the human *telos*, has for modern men and women become one of self-management and/or self-construction. The human task is no longer understood in terms of fitting into the nature of things, but is now understood to entail knowing and utilizing natural objects—including one's body—in the service of freely chosen purposes. "Descartes calls on the enquiring mind to disengage from the cosmos," Taylor writes, "but we have to disengage also from our own bodily nature, not just because the body is another part of the 'external' universe, but also because in order to achieve the proper rational mastery, the soul must instrumentalize the passions."[16]

Descartes is often called the father of modern philosophy, and it is not difficult to see why. His new conceptualization of nature would prove foundational to the development of the modern scientific method, a method that has permitted us to achieve a great deal of power over nature. A latent Cartesianism is evident in a great many other aspects of modern civilization as well, and not just in philosophy, but in methods of administration and military organization, in politics, in education, and in the modern ideal of the individual agent who is able to remake him- or herself by means of methods and techniques arising out of disciplined scientific knowledge. Indeed, modern civilization all but requires us to take an instrumental stance vis-à-vis the natural world, including our own bodies. As Taylor comments, we moderns tend to believe that our desires, inclinations, tendencies, habits of thought and feeling must be "worked on" until they conform to our desired specifications.[17]

Yet the Cartesian disconnection of disembodied "subjects" from the world of natural "objects"—a world that includes our bodies—has posed a conundrum for modern thought. How then are we to understand the relations of mind and body, of ourselves to the natural world? Poteat

---

[15]Charles Taylor, *Sources of the Self: The Making of the Modern Identity* (Cambridge, MA: Harvard University Press, 1989), 615.
[16]Ibid., 611-12.
[17]Ibid., 612-13.

argued that the Cartesian bifurcation of thinking substances from the lifeless stuff of the world has seriously prejudiced the modern imagination, rendering it insensitive to temporality and place, and leaving it susceptible to gnosticism. "No one starting with the same picture of man's relation to nature," he noted,

> and therefore being constrained by the logic of this picture, has been able to offer, in the three hundred years since Descartes, a more plausible suggestion as to how minds and bodies are to be brought together. During this time hardly any of our debates, explicit or by implication, over the nature of man in the world has failed to move, rather despairingly, back and forth between matter and mind, object and subject, the world as we cannot help fancying it to be, eternally in itself, and the world as it presents itself to us in our perceptions.[18]

The impact of Cartesianism upon the modern imagination has also been, as Alexandre Koyré noted, to tear us out of a finite and well-ordered world—out of what had once been considered a "hierarchy of perfection and value"[19]—and to cast us into a frightfully infinite "space" now held together only by the uniformity of its basic elements and laws. Putting this somewhat differently, Cartesianism plucked us out of a divinely created *cosmos* or order, in which we had been created to fit, and thrust us into a homogeneous universe of matter in motion that is apparently indifferent to human interests. Within this universe, we must carve out space for ourselves by means of science and technology. As philosopher Robert Doede notes,

> The warm and cozy hierarchical medieval cosmos, where nature was immanently animated by the desire to attain certain goals (final causes), was replaced with the frightfully infinite and homogenous uni-verse of space and inanimate matter, mechanically churning out consequences. . . . consequences that, by functionally correlating them with the algorithms of certain mathematical formalisms, became predictable, and thus exploitable, by the impositions of the human will.[20]

---

[18]Poteat, *Polanyian Meditations*, 253.

[19]Alexandre Koyré, *From the Classical World to the Infinite Universe* (New York: Harper Torchbooks, 1957), vi.

[20]Robert Doede, "The Decline of Anthropomorphic Explanation: From Animism to Deconstructionism," unpublished paper delivered to the Regent College faculty, Vancouver, Canada, 1992, 7.

Having stripped nature of everything but extension, Doede continues, human agents were left free, being *supra*-natural, to impose their own final causes upon it. This is the theoretical substructure of modern technology: employing impersonal and mechanical matter to realize humanly willed purposes. Doede notes that such a view of the world, and of human agents' central place within it, first became thinkable within Descartes's philosophical method, a method that then initiated a kind of inexorable progression in the natural sciences: objectify—unitize—mathematize—technologize.[21] Mumford emphasizes similarly, "To fix attention upon a mechanical system was the first step toward creating system: an important victory for rational [scientific] thought."[22]

Yet the success of Cartesianism has also blinded us from imagining other ways of being-in-the-world. In denying final purposes to nature, it has also denied them to us. In addition, although life can be understood as a kind of mechanism, mechanisms cannot be said to be alive. Again quoting Mumford, "Unfortunately, isolation and abstraction, while important to orderly research and refined symbolic representation, are likewise conditions under which real organisms die."[23] What is thus left to Cartesian understanding, Mumford laments, is often simply a depopulated world of matter in motion, a "wasteland."[24]

The Cartesian wasteland is, however, one in which human engineers have become extraordinarily powerful. After all, modern scientific understanding has enabled us to penetrate nature and to unleash the power of steam, electricity, chemicals, and even atomic energies. Science has enabled us to decode the genome, revealing the chemical information that regulates and directs the life process. Something also seems to have been unleashed within us, corresponding to this rationally liberated energy and knowledge. As Guardini observed, "A specific attitude, craving, approach, a desire for mechanical and rational works—has arisen and placed these forces at our disposal, creating for them the intellectual plane on which we can see and exploit them with

---

[21]Ibid., 9.
[22]Mumford, *Technics and Civilization*, 47.
[23]Ibid., 50.
[24]Ibid., 51.

increasing fullness."[25] Philosopher Arnold Gehlen sought to summarize the essence of the modern technical spirit in a term borrowed from Max Scheler: *pleonexia*, simultaneously signifying greed, arrogance, and ambition for power. Pleonexia, Gehlen believed, best captures the social psychology of the modern technological age.[26]

As significant as Cartesian philosophy was for the modern technological outlook, however, Descartes appears only to have made explicit many assumptions that were implicit already in a way of understanding the human task within the world that dates back to the end of the Middle Ages. The distinctively modern drive to master and to possess nature—including the human body—as well as the fact that the rational mastery and possession of nature continue to be pursued very largely "in good conscience," appear at least in part to be legacies of the Protestant Reformation.

Protestantism and the Duty of Mastery

The characteristically modern, and now postmodern presumption that we are—or at least should be—largely free to make and remake ourselves by methodical and disciplined action, making use of modern science and technology, was not initially nihilistic. Neither did it stem at first from Cartesian self-objectification. It appears rather to have been a kind of unintended outgrowth of a distinctively Protestant understanding of religious duty.[27] Enthusiasm for disciplined empirical reasoning in the practical reform of material life was, for example, one of the distinctive features of seventeenth-century English Protestant culture. This was the gist of American sociologist Robert K. Merton's seminal article, "Motive Forces of the New Science."[28] Taking as his point of departure Max Weber's earlier but unsubstantiated suggestion that the practical Protestant ethic had been just as critical to the rise of

---

[25]Romano Guardini, *Letters from Lake Como: Explorations in Technology and the Human Race* (Grand Rapids: Eerdmans, 1994 [1923]), 71-72.

[26]Arnold Gehlen, *Man in the Age of Technology*, trans. Patricia Lipscomb (New York: Columbia University Press, 1980), 109.

[27]Portions of the following section have been excerpted from my *The Way of the (Modern) World: Or, Why It's Tempting to Live as if God Doesn't Exist* (Grand Rapids: Eerdmans, 1998), 110-13.

[28]Robert K. Merton, "Motive Forces of the New Science," in *Puritanism and the Rise of Modern Science: The Merton Thesis*, ed. I. Bernard Cohen (New Brunswick, NJ: Rutgers University Press, 1990 [1938]), 112-31.

early-modern science as it had been to the development of capitalism, Merton notes that English Puritanism and the "new science" did indeed share a number of basic assumptions, and that this goes some distance toward explaining why early-modern scientific understanding enjoyed such popularity and was so rapidly diffused within seventeenth-century English society. Among these shared assumptions were the importance of improving the material living conditions of ordinary people, the value of disciplined and careful reasoning, and a willingness to subject assertions to empirical verification. The significance of these shared assumptions, Merton contends,

> is profound though it could hardly have been consciously recognized by those whom it influenced: religion had, for whatever reasons, adopted a cast of thought which was essentially that of science so that there was a reinforcement of the typically scientific attitude of the period. This society was permeated with attitudes toward natural phenomena which were derived from both science and religion and which unwittingly enhanced the continued prevalence of conceptions characteristic of the new science.[29]

A number of qualifications were subsequently offered to Merton's original thesis. In *Religion and the Rise of Modern Science*, Reijer Hooykaas takes issue with Merton's suggestion that the affinities between Puritanism and early-modern science were only coincidental. On the contrary, Hooykaas argues, the consonance between the two movements must surely have been recognized by both Protestant theologians and early scientific researchers. The similarity between the Puritan suspicion of ecclesiastical authority, for example, and early-modern science's refusal to trust medieval and classical authorities was not coincidental. Both followed from Protestant convictions about authority and the priesthood of all believers before God. On the one hand, defenders of the new science "were perfectly conscious of the analogy linking the liberation from ecclesiastical and philosophical tradition by the Reformation and the liberation of science from ancient authority by the new learning."[30] The Puritan divines, on the other hand, appear to have become convinced that the liberation of science and

---

[29]Ibid., 131.
[30]Reijer Hooykaas, *Religion and the Rise of Modern Science* (Grand Rapids: Eerdmans, 1972), 113.

theology from scholastic rationalism could not help but benefit the religious reformation of English society.

Additional qualifications to the Merton thesis were elaborated by Charles Webster in a historical study entitled *The Great Instauration: Science, Medicine, and Reform, 1626–1670*.[31] While agreeing with Merton, Hooykaas, and others who stressed the relevance for science of Puritanism's practical and utilitarian bent, Webster also draws attention to the significance of Puritanism's commitment to education, and particularly to specialized vocational education. Webster also points out the Puritan proclivity to found organizations or "societies" dedicated to specific practical purposes. It was precisely this activistic practicality, Webster believes, that accounted for the ready acceptance of experimental science in seventeenth-century England:

> The Puritans were dedicated to unremitting exertion, and increasingly they were sympathetic to the virtues of manual labor; successful accomplishment of works provided one of the major means whereby some intimation of election might be obtained, and the sciences, particularly the utilitarian sciences, were one of the avenues whereby substantial works could be performed. The Calvinist God was distant and inscrutable, but the patient and accurate methods of experimental science, penetrating slowly towards the understanding of the secondary causes of things in the search for a gradual conquest of nature, represented the form of intellectual and practical endeavor most suited to the Puritan mentality. Immediate goals were conceived with proper humility, and progress was necessarily slow, but every step brought further insight into the providence of God, so constantly reaffirming the correctness of the procedure.[32]

As Webster's observations suggest, Puritan eschatology also appears to have contributed to the legitimation of early modern scientific investigations. The practical conquest of nature, they believed, furthered the recovery of the human dominion over creation that had been lost at the fall. Each incremental step in the subjugation of nature was believed to be a step in the direction of the New Jerusalem.[33]

---

[31]Charles Webster, *The Great Instauration: Science, Medicine, and Reform, 1626–1670* (New York: Holmes & Meier, 1976).
[32]Charles Webster, "'Conclusions' to The Great Instauration," in *Puritanism and the Rise of Modern Science*, ed. I. Bernard Cohen (New Brunswick, NJ: Rutgers University Press, 1990), 282.
[33]Ibid., 283.

The technical, instrumental, and scientific cast of modern civilization, in short, appears initially to have been given decisive impetus by the practical activism of Calvinist Protestantism and its determination, in effect, to reengineer both the church and the world. What this meant, Taylor observes, was that the instrumental stance toward the world had been given a new and important spiritual meaning:

> It is not only the stance which allows us to experiment and thus obtain valid scientific results. It is not only the stance which gives us rational control over ourselves and our world. In this religious tradition, it is the way we serve God in creation. And that in two respects: first, it is the stance we must assume to work in our callings to preserve ourselves and God's order; but second, it is also what protects us against the absorption in things which would wrench us away from God. We must constantly remember to treat the things of creation merely as instruments and not as ends valuable in themselves. Richard Sibbes enjoins us to "use [the world] as a servant all thy dayes, and not as a Master"; and he tells us. . . . "Labour therefore to have the world in its owne place, under thy feet." Instrumentalizing things is the spiritually essential step.[34]

Of course, the practical conquest of nature eventually ceased to be understood within a Christian theological and ethical framework and as a religious duty. And in the absence of religious understanding, a kind of nihilism did begin to surface within North American culture, a nihilism that has by now, together with the creative-destructive spirit of late-modern capitalism, become a hegemonic cultural force. This is the first half of George Grant's provocative thesis in "In Defense of North America."[35] The other half of Grant's argument is that, while the practical conquest of nature is no longer sought for the sake of the kingdom of God, it is still pursued with a kind of quasi-religious intensity. For it has come to be believed that modern technological development will eventually give rise to a kind of rationalized "kingdom of man."[36] It might seem, Grant observes, that Calvinist practicality has devolved simply into the drive to technological development for its own sake. Yet "what makes the drive to technology so strong

---

[34]Taylor, *Sources of the Self*, 232.
[35]George Grant, "In Defense of North America," in *Technology and Empire: Perspectives on North America* (Toronto: Anansi, 1969), 15-40.
[36]Ibid., 25.

is precisely that it is carried on by men who still identify what they are doing with the liberation of mankind."[37] In this connection, Grant notes that this kind of secularized faith in technological liberation is shared across the political spectrum. Indeed, he believes that the technological imperative has all but eliminated any human desiring within North American culture beyond the desire to make the future by means of technological mastery. The technological imperative has all but closed down any other kind of thinking beyond that of practical-rational calculation. "It is in this sense," Grant laments, "that it has been truthfully said: technology is the ontology of the age. Western peoples (and perhaps soon all peoples) take themselves as subjects confronting otherness as objects—objects lying as raw material at the disposal of knowing and making subjects."[38]

Defending modern technological development in the name of "progress" has fallen out of favor since Grant wrote in the late 1960s. Still, modern technology is championed under the banner of human liberation, and events like Apple Computer's annual staged product announcements have become quasi-religious rituals for many people. To the extent, then, that we continue to confront our world as a series of problems for which technical solutions must now be developed, this is not simply the legacy of Cartesianism but also that of the Protestant conviction that the kingdom of God was to be ushered in by way of the practical reformation of society and culture.

Late-Medieval Developments

Yet our quest for the origins of the mechanical world picture needs to be extended back before the Reformation. After all, Protestant willingness to manipulate both nature and culture for the sake of imposing religious purposes upon them, from which Protestant enthusiasm for the potential of early-modern science and technology appears to have stemmed, emerged out of a theological milieu that had already begun to question the older understanding of the natural world as a divinely ordained hierarchy of perfection and value. Doubts had long since surfaced about the classical assumption that the human task was simply to fit into "the nature of things."

---

[37]Ibid., 27.
[38]George Grant, *Technology and Justice* (Toronto: Anansi, 1986), 32.

Indeed, the origins of the distinctively modern world picture may be traced back at least as far as theological innovations that took place in the thirteenth and fourteenth centuries in Europe, innovations that even at the time were referred to as the *via moderna*, or "modern way."

Before the *via moderna* arose, an older classical tradition, known as the *via antiqua*, had long stressed that the order of Nature (note: in uppercase) was a "Good" wholly independent of human willing, into which human beings must somehow fit if they were to properly realize their *teloi*. A life well lived, from this perspective, entailed striving by means of disciplined and discursive reasoning to discern "the Good" and, once it had been discerned, to strive to live in accordance with it. Science, on this classical account, was primarily deductive and not empirical, for since Plato it had been assumed that the natural order owed its existence to and in a sense participated in a larger and ultimately divine order available finally only to theoretical reason. True knowledge of Nature was not to be sought by way of observation; rather, it was to be inferred from the descriptions and principles found in authoritative philosophical and theological texts. It was assumed that the natural order of things represented a kind of dim reflection of a heavenly order that our souls would enter after death. The label commonly attached to this largely Platonic understanding of Nature was *realism*.

Aristotle modified Plato's philosophy of Nature by contending that we need not actually decide whether and where there is a realm in which ultimately real things exist over and against those realities we actually experience. Neither is it necessary to contend that universal "Forms" somehow inhere within their individual empirical representatives. Instead, Aristotle simply insisted that universal principles reveal themselves when individual representatives of said principles are multiplied in our experience. Aristotelian science was, therefore, empirical up to the point of abstracting the essential qualities from sensible particulars. Once these essential qualities had been identified, however, science was to proceed largely deductively by way of syllogism and logical demonstration.

Aristotelian understanding of Nature was also *teleological*, concerned with discerning the ultimate purposes of things. Indeed, it is for this reason that Aristotelian science is often described as anthropomorphic, for Nature is understood on the analogy of concerns and priorities that we find within ourselves, such as our propensity to pursue the question "why?" all the way to the final purpose of things. Aristotle's modification of Plato is sometimes called moderate or immanent realism.

Within the biblical tradition, similarly, it had long been stressed that the human creature is subject to divine authority as revealed in Scripture, and that we are embedded within a creation that is not of our own making—and to which we are also subject. The "fear of the LORD" that is the beginning of wisdom (Prov 9:10) entails meditation on the law of the Lord (Ps 1:2) as well as admiration for God's marvelous works in creation (Ps 104). While the ontological foundations for a genuinely empirical science may well have been present within the biblical tradition in the affirmation of God's goodness and in the doctrine of creation *ex nihilo*,[39] the authors of Scripture do not speculate about "nature" per se. As philosopher and theologian Diogenes Allen noted, they are not led to posit a Creator out of their desire to understand and explain the natural world, but rather the reverse is the case. The authors of Scripture celebrate created nature because it bespeaks God's goodness and faithfulness. Their belief in God, furthermore, is not derived from the order they discern in the natural world, but rather it rests upon God's revelation of himself to Abraham, Isaac, and Jacob,[40] and ultimately in the person and work of Jesus Christ.

The goal of human life from within the biblical tradition, then, was to qualify for "the life to come" (1 Tim 4:8) by knowing the God who had so graciously seen fit to reveal himself to us. And while human creatures, having been created after the "image" and "likeness" of God, had been enjoined to "have dominion over the fish of the sea and over the birds of the heavens and over every living thing that moves on the earth" (Gen 1:28), this dominion would never have been construed to mean the imposition of merely human purposes upon natural stuff. Rather, the restoration of human dominion over nature is something for which we must patiently wait. Just as created nature is said to wait in anxious anticipation for the revelation of

---

[39]See Michael B. Foster, "The Christian Doctrine of Creation and the Rise of Modern Natural Science," *Mind* 43 (1934): 446-68.

[40]Diogenes Allen, *Philosophy for Understanding Theology* (Atlanta: John Knox, 1985), 3.

the "children of God" (cf. Rom 8:19), so we must also wait for that day on which we will be "clothed . . . with immortality" (1 Cor 15:52-54).

The assimilation of the classical philosophical science of Nature—principally of Aristotle's immanent realism—into Christian theology was the crowning intellectual achievement of the Christian Middle Ages. As theologian Ernst Troeltsch detailed in *Protestantism and Progress: The Significance of Protestantism for the Rise of the Modern World*, this synthesis was, in effect, the last great effort of antiquity:

> The immediate intrusion into the world of the Divine, with its laws, forces, and ends, exactly definable against the background of purely natural capacity, determines everything, and produces an ideal of civilisation which, in theory at least, signifies a direction of mankind as a whole by the Church and its authority—an ideal which everywhere authoritatively determines the mode of combination of supernatural, Divine, with natural, earthly, human ends. Supreme over all is *Lex Dei*, which is composed of the *Lex Mosis* or the Decalogue in combination with the *Lex Christi* and the *Lex Ecclesiae*, but also as *Lex Naturae* includes within itself the juridico-ethical and the scientific heritage of antiquity, and the natural claims of life.[41]

The theory that regulated this grand synthesis, Troeltsch continued, is that both sets of laws, the biblical/ecclesiastical and the natural, are ultimately the same, since they originally coincided. "It is only now," he wrote, "and for sinful humanity, that they diverge, and under the direction of the Church their proper equivalence will again be restored, though now, indeed, conditioned by the continuance of original sin."[42]

Platonic realism had long posed any number of intellectual difficulties, however. Where, after all, could this putative realm of "universals" be said to be? And in what sense could such a realm be more real than the reality we ordinarily experience? Serious doubts began to surface toward the end of the Middle Ages even in respect to Aristotle's immanent realism. After all, if universals only actually exist in our experience of particulars, then aren't they simply names for the patterns we discern in our experience? If, furthermore,

---

[41]Ernst Troeltsch, *Protestantism and Progress: The Significance of Protestantism for the Rise of the Modern World* (Philadelphia: Fortress, 1986), 21-22.
[42]Ibid.

there is no realm in which universals may be said to exist, as Aristotle had averred, then where can God be said to exist? Concerning this latter question, thirteenth-century scholar John Duns Scotus (1266–1308) contended that reality is all of a piece and therefore that God's existence per se does not differ from that of everything else that exists. Being, on Scotus's account, is univocal. God, in effect, is "mappable on the same set of coordinates as creatures."[43] With respect to our actual knowledge of God's creatures, furthermore, the Franciscan philosopher William of Ockham (d. 1349) maintained—over and against Scholastic rationalists who continued to insist that the meanings and purposes of created things could be deduced from their essences, as well as from theological first principles—that whatever nature may be said to mean, its meaning is entirely immanent within it and discernible only by way of careful observation and experience.

Ockham's "modern way," as it came to be called, had a number of profound implications, though they would take several centuries to work themselves out within early-modern understanding. In the first instance, the *via moderna* demoted medieval natural theology, which had come to be regarded as the acme of natural knowledge, to the simple observation of nature (note: now a lowercase "n"). Ockham's modern way also had the effect of restricting the notion of causality in nature to efficient causality: (A) may be said to cause (B) if (B) regularly and repeatedly follows (A) and if, should (A) not occur, (B) does not follow. Just why the one should "cause" the other, furthermore, is explicable in largely mechanical terms. The Aristotelian notion of "final" causation, whereby nature was thought to be moved by something akin to desire from potency to act, was regarded by Ockham as merely metaphorical.[44] In Ockhamist understanding, as Allen commented,

> There is no possibility of inferring from a knowledge of one thing that something else must result from it, as is the case with essences, which once abstracted permit one to go beyond the particulars that have been experienced. Ockham's view of causality thus makes demonstration impossible, for our

---

[43]Robert Barron, cited in Brad Gregory, *The Unintended Reformation: How a Religious Revolution Secularized Society* (Cambridge, MA: Harvard University Press, 2012), 37.

[44]See Andre Goddu, "Ockham's Philosophy of Nature," in *The Cambridge Companion to Ockham*, ed. Paul Vincent Spade (Cambridge: Cambridge University Press, 1999), 154.

knowledge can extend no further than our experience. Beyond our experience there is only probability.[45]

Probability, while not carrying the certainty of rational demonstration, does at least have the advantage of being mathematically calculable, as well as corroborating our actual experience of natural occurrences. If nature is reliable, furthermore, this is not because it is rationally necessary, but because a loving and faithful God has willed and continues to will nature to operate in an orderly way.

In Ockhamist theology, God was not bound to have created nature in the ways that he actually created it. If, then, we would arrive at a true knowledge of created nature, we must take care to observe what God has actually chosen to make. In thus encouraging a genuinely empirical investigation of nature, Ockham's *via moderna* may be said to have provided a kind of onto-theological foundation for the development of early-modern science.[46] In addition, and as noted above, late-medieval nominalism (as Ockham's position was also called) came to have a certain resonance within Protestantism. The Reformers were also concerned to stress that God did not have to redeem his creation—including us—in the manner that he actually has, which is precisely why grace is so amazing. The *via moderna* also proved useful to the Protestant defense of *sola Scriptura* over and against Scholastic rationalism, for, as Calvin in particular would stress, the Scriptures must be allowed to speak for themselves. The critical question was not, "What can be

---

[45] Allen, *Philosophy for Understanding Theology*, 155.

[46] Though this influence extended only so far, for as Oliver O'Donovan has pointed out, the notion that generic equivalence simply reflected the imposition of mind upon a universe of particulars was at odds with early-modern science's concern to discover patterns in nature for explanation and for reliable prediction. "[I]n the long run," O'Donovan noted, "science found nominalism an enemy to its project; for science is interested in nothing if not regularities, and nominalism must deny that the regularities which science purports to observe are real. Scientific endeavor has been sustained not only by the sense that it may *dare* to think new thoughts about the world order (important as that has been), but also by the conviction that its thoughts, when supported experimentally, can actually declare (with whatever limitations) something of the character of objective reality. The denial of real regularities could never be more than a goad, driving it to discard the obvious and straightforward relationships in search of hidden structures of greater complexity. If the denial were taken absolutely seriously, science would lose its peculiar *raison d'être* and be absorbed once again into the category of simple speculation from which it once sprang forth." Oliver O'Donovan, *Resurrection and Moral Order: An Outline for Evangelical Ethics* (Grand Rapids: Eerdmans, 1986), 48.

said about the Scriptural text on the basis of what we already know, say, on the basis of what the church has already authoritatively established?" but rather, "What, upon close examination, do the texts actually disclose?" In this connection, theologian Thomas Torrance argued that it was Calvin's careful and inductive approach to Scripture that provided the model for early-modern empirical science:

> We know God by looking at God, by attending to the steps he has taken in manifesting himself to us and thinking of him in accordance with his divine nature. But we know the world by looking at the world, by attending to the ways in which it becomes disclosed to us out of itself and thinking of it in accordance with its creaturely nature. Thus, scientific method began to take shape both in the field of natural science and in the field of divine science.[47]

Yet late-medieval nominalism also appears—at least we can say this in retrospect—to have overemphasized the relative autonomy of created nature. It also appears to have prepared the way for a dangerous misunderstanding of the place of human beings within this relatively autonomous natural world. For in its rejection of teleology, the *via moderna* suggested that the observing mind simply encounters a kind of inert creation. "Not a creation without movement," O'Donovan observed,

> but a creation without a point to its movement. Thus the mind credits to its own conceptual creativity that teleological order which is, despite everything, necessary to life. All ordering becomes deliberative ordering, and scientific observation, failing as it does to report given teleological order within nature, becomes the servant of *techne*. Of course, man continues to eat vegetables; but he no longer knows that he does so because vegetables *are* food, and comes to imagine that he has *devised* a use for them as food.[48]

The *via moderna* may also be said to have inverted the traditional relation between theology and experience. Instead of preceding and preparing us to understand our experience of the world, theology was increasingly left to follow behind observation and experience, making sense of both as best it could. As Tage Lindbom noted concerning Ockham's contribution to secular modernity,

---

[47]Thomas F. Torrance, *God and Rationality* (London: Oxford University Press, 1971), 39.
[48]O'Donovan, *Resurrection and Moral Order*, 52 (emphases added).

For traditional [religious] man . . . reality is a translucent objectivity, perceptible through the world of sensory experience; natural things are lamps through which shines the light of their heavenly—ultimately Divine—prototypes. But Ockham replaces this with . . . the consciousness that we gain through our sensory and mental faculties. The real is the evidence of the senses. Ockhamism informs "truth" or reality with a new content: it is this-worldly, subjective, [and] gained by mental processes, especially rational and logical processes that depend on sense data.[49]

Late-medieval nominalism is thus sometimes credited with setting the stage for what would become an increasingly profane conception of human existence. It would eventually occur to modern thinkers, so the argument runs, that once created nature can no longer be reasoned theologically, teleologically, and in terms of its participation in a heavenly realm, then neither can human nature.[50] Indeed, even the belief in God would come to many modern thinkers to seem irrelevant to life in this world. In this connection, the preponderance today of the notion that it is up to us to ascribe meaning and significance to the natural world is a good indication that this was, in fact, the way at least some of the heirs to the Reformation did go on to interpret the *via moderna*. Theologian Colin Gunton commented:

God was no longer needed to account for the coherence and meaning of the world, so that the seat of rationality and meaning became not the world, but the human reason and will, which thus displace God or the world. When the unifying will of God becomes redundant, or is rejected for a variety of moral, rational and scientific reasons, the focus of the unity of things becomes the unifying rational mind.[51]

An arcane late-medieval theological disputation over the nature of reality may thus lie behind the activistic, manipulative, anthropocentric, and very largely secular bent of modern technological culture. Modern science's understanding of nature, after all, does not simply disclose the fact that nature has been disenchanted, but also that nature is theologically mute. We simply

---

[49]Tage Lindbom, *The Myth of Democracy* (Grand Rapids: Eerdmans, 1996), 20.

[50]See, for example, Hans Boersma, *Heavenly Participation: The Weaving of a Sacramental Tapestry* (Grand Rapids: Eerdmans, 2011).

[51]Colin E. Gunton, *The One, the Three and the Many: God, Creation and the Culture of Modernity* (Cambridge: Cambridge University Press, 1993), 28.

do not expect it to disclose any kind of metaphysical truth beyond, perhaps, the notion that the agency—if, indeed, there is such a thing—responsible for natural order is creative and powerful. This narrowing of nature's theological significance appears to have created a great deal of space for scientific inquiry as well as for the technological manipulation of natural stuff. Yet it also contributed substantially to the sterility, barrenness, and impersonality of modern technological society and culture. As historian Brad Gregory put it recently, "Desacramentalized and denuded of God's presence via metaphysical univocity and Occam's razor, the natural world would cease to be either the Catholic theater of God's grace or the playground of Satan as Luther's *princeps mundi*. Instead, it would become so much raw material awaiting the imprint of human desires."[52]

The Eclipse of "Being" Within the Western Philosophical Tradition

It has been suggested that the genealogy of the mechanical world picture extends back even beyond the Christian Middle Ages. In a celebrated short essay entitled "The Question Concerning Technology," philosopher Martin Heidegger contended that the supposition that we are free to impose our will upon the raw stuff of nature antedates Christianity by centuries, and that it runs right through the Western philosophical tradition.[53] While a generalized awareness that humans might master and possess nature (for good and for ill) has only surfaced within the last hundred and fifty years or so, this is simply because our technologies have become so much more powerful and have been augmented by modern scientific understanding. Heidegger was concerned to emphasize, however, that technological development in the West has been tending toward the mastery and possession of nature for a very long time, indeed perhaps since the time of Plato and Aristotle.

Human beings, Heidegger believes, are intrinsically and necessarily creative, continually producing and reproducing "worlds" in which to live and work. Human "world making" is manifest in language as well as in the

---

[52]Gregory, *Unintended Reformation*, 57.
[53]Martin Heidegger, "The Question Concerning Technology," in *The Question Concerning Technology and Other Essays*, trans. William Lovitt (New York: Harper & Row, 1977), 3-35.

making and using of tools. Although technology and poetry are typically thought to be very different kinds of activities, Heidegger stressed that both are aspects of human *poiesis*, that is, of making and of "bringing forth." Both reveal a human "world" and disclose suppositions concerning human beings' relation to given nature.

While human beings have, so far as we know, always revealed and shaped their worlds by developing and using various tools, Heidegger believes that modern technological development is essentially different from traditional tool making. The purpose of his relentless questioning in "The Question Concerning Technology" is to try to uncover this essential difference. The key, he suspects, lies in modern technology's "challenging" and "setting upon" given nature with the goal of setting things apart for future use.[54]

"But," one might interject, "hasn't this always been true of human making? Haven't people always used nature? Isn't this true, for example, of a windmill?" "No," Heidegger replies in a manner reminiscent of the distinction Guardini made between contrivances and machines, because the windmill depends upon the blowing wind. It does not extract energy from the air in order to store it for future use.[55] By contrast, modern technology unlocks and isolates forces previously hidden within nature and transforms them into "resources" that can be stored and then used at will. "Unlocking," "transforming," "storing," "distributing," and "switching about," Heidegger contends, most certainly represent a kind of human making. They do indeed reveal a definite relation to given nature, but this violent kind of making reveals a relation to nature that is both restless and relentless. It discloses a relation to nature that seeks to regulate and to secure but that can never arrive at regulation or security.

Whatever is "seen" and/or known in this new modern way, Heidegger suggests, becomes "standing-reserve."[56] The crucial characteristic of things that have been thus revealed as standing-reserve—of things that have, to use Heidegger's phrasing, been "made present"—is that they no longer stand over and against us as things with their own integrity, apart from our use of

---

[54]Ibid., 16.
[55]Ibid., 14.
[56]Ibid., 17.

them. Instead they have simply been converted into resources that "stand by" to be used for whatever purposes we might eventually devise for them.

Ironically, human beings are themselves subject to this same conversion process. We have also become standing-reserve within the modern purview—as in the expression "human resources." This perplexing feature of modern self-understanding relates directly to our discussion of Cartesian self-objectification.

Heidegger labels this peculiarly modern way of world-making or, as he puts it, this "challenging-forth" of given nature that reveals all things—including people—to be standing-reserve, as "enframing."[57] It is a way of "seeing" the world and ourselves, and of envisioning human purposes in the world. This new way of seeing the world, he believes, and this new way of envisioning our purposes within the world, is the essence of modern technology. It is not itself a technology, but it has made modern technological development possible by revealing a world in which all things stand ready to be used for purposes that we are free to choose. "The essence of modern technology," Heidegger writes, "starts man upon the way of that revealing through which the real everywhere, more or less distinctly, becomes standing reserve."[58]

Enframing, Heidegger continues, is not simply the essence of technological making; it is also the essence of scientific knowing. "Modern physics," he writes,

> is not experimental physics because it applies apparatus to the questioning of nature. Rather the reverse is true. Because physics, indeed already as pure theory, sets nature up to exhibit itself as a coherence of forces calculable in advance, it therefore orders its experiments precisely for the purpose of asking whether and how nature reports itself when set up in this way.[59]

Put somewhat differently, modern science does not enter into a dialogue with nature so much as it interrogates it. It confronts nature from a particular point of view and with a particular result in mind. In effect, modern science insists that nature reveal herself in terms of elements and processes that are regular, repeatable, and mathematically representable. Only when

---

[57]Ibid., 20.
[58]Ibid., 24.
[59]Ibid., 21.

nature reports itself in this way can it be said to be known "scientifically." It must be known scientifically, furthermore, to become available for use. In this connection, while it may well have taken decades, even centuries, for the essence of modern technology to appear in actual technological achievements, enframing was already present within early-modern science's distinctive way of interrogating nature. "Chronologically speaking," Heidegger writes, "modern physical science begins in the seventeenth century. In contrast, machine-power technology develops only in the second half of the eighteenth century. But modern technology, which for chronological reckoning is the later is, from the point of view of the essence holding sway within it, the historically earlier."[60]

Has the new modern way of knowing and making become inexorable and irresistible? Has it become our fate? "No," Heidegger avers, but it has placed us in a perilous position, for now it is difficult for us even to imagine, much less to appreciate, any other way of knowing and making—indeed, any other way of being-in-the-world. Again, as Grant (following Heidegger) put it so succinctly, "The very substance of our existing which has made us the leaders in technique, stands as a barrier to any thinking which might be able to comprehend technique from beyond its own dynamism."[61] Such is our situation, and it is evident in two important respects: in the first instance, having now become the masters and possessors of nature, we find that nature, including human nature, has been overwhelmed by human making. Not only are we increasingly surrounded by humanly constructed technological artifacts, but whatever nature remains has now been revealed by modern science as an "environment" comprised of "natural resources" that may either be used or stewarded as we see fit. Nothing appears to have any inherent integrity apart from whatever use we might devise for it. "The impression comes to prevail," Heidegger comments in this connection, "that everything man encounters exists only insofar as it is his construct. This illusion gives rise in turn to one final delusion: It seems as though man everywhere and always encounters only himself."[62]

---

[60]Ibid., 22.
[61]Grant, "In Defense of North America," 40.
[62]Heidegger, "The Question Concerning Technology," 27.

This, Heidegger contends, is a delusion, because enframing actually prevents us from truly knowing ourselves. "*In truth,*" he emphasizes,

> precisely nowhere does man today any longer encounter himself, i.e., his essence. Man stands so decisively in attendance on the challenging-forth of Enframing that he does not apprehend Enframing as a claim, that he fails to see himself as the one spoken to, and hence also fails in every way to hear in what respect he ek-sists [*sic!*] from out of his essence, in the realm of an exhortation or address, and thus can never encounter only himself.[63]

Putting Heidegger's important point somewhat differently, our knowledge of ourselves is never direct. It is reflected back to us in the relations that we have with "others," which includes our lived environment. If these relations are stunted and/or distorted, they will reflect back an image of ourselves that is also stunted and/or distorted. Here we might recall G. W. F. Hegel's celebrated master-slave dialectic in which *both* master and slave are prevented, given the brutality of the master-slave relation, from realizing their full humanity.[64] Given the brutality of our distinctively modern relation to given nature, Heidegger suggests along similar lines, it is hardly surprising that we moderns struggle with self-knowledge. Enframing inhibits us from truly listening to given nature and thereby prevents us from experiencing the existential change that might have come through our appreciation of its otherness. In failing to listen to nature, we foreclose upon the possibility of discerning who we could have become in relation to it.

Here we might recall Martin Buber's celebrated analysis of the "I-It" conjunction as over against the "I-Thou" relation.[65] A more common term for enframing, after all, is "objectification," the reduction of others to mere objects. The occasion of Buber's celebrated little book *I and Thou* was his concern that modern scientific understanding had encouraged us to

---

[63]Ibid. (emphasis in original).

[64]See G. W. F. Hegel, "The Independence and Dependence of Self-Consciousness: Lordship and Bondage," in *Phenomenology of Spirit*, trans. A. V. Miller, with analysis of the text and foreword by J. N. Findlay (New York: Oxford University Press, 1977), 312.

[65]See Martin Buber, *I and Thou*, trans. Walter Kaufmann (New York: Charles Scribner's Sons, 1970 [1937]). I represented Buber's reasoning in a very similar way in my earlier work *The Way of the (Modern) World*, 296-301.

objectify both the world and one another. Yet the act of objectification—of establishing the I-It relation—does not involve one's whole self, and neither does it initiate a genuine conversation. Rather, it discloses a one-sided relation in which there is really only one active voice. Objectification does not expose the self, therefore, to a possibility of relational mutuality. It fails to yield genuine self-knowledge and the possibility of personal growth. As Alistair McFadyen noted,

> Intending someone or something as an object is to intend the relation as a monologue. For an object is intended and perceived as having no independent meaning or existence apart from this relation. It cannot offer a point of moral resistance because it is not perceived as ethically transcendent. The relation can only be exploitative and manipulative. . . . The I of an I-It relation has an unbounded sense of its proper claims, seeking from the other only that which is a confirmatory repetition of itself. In seeking oneself from the other, one is engaged in one-way communication open only to oneself.[66]

This "one-way communication open only to oneself" does have certain advantages, for it augments our ability to control the world in many ways. Yet objectification also inhibits our ability to enter into an I-Thou relationship with the world, with other people, and ultimately with God. The objective attitude makes it difficult, for example, to apprehend beauty; for anything that is only valued as a resource to be utilized cannot really be acknowledged as beautiful, at least not in the classical sense of beauty. The objective spirit is also obviously destructive in and of interpersonal relationships. While the objective utilization of others may enable us to "get things done," it is not conducive to family, friendship, camaraderie, and fellowship. Neither is objectification conducive to genuine worship, for the living God simply does not allow himself to be known objectively, as an object of human manipulation. If we seek to know God in this way, he retreats from our view. As Buber noted,

> It becomes necessary to proclaim that God is "dead." Actually, this proclamation means only that man has become incapable of apprehending a reality absolutely independent of himself and of having a relation with it. . . . Man's

---

[66]Alistair I. McFadyen, *The Call to Personhood: A Christian Theory of the Individual in Social Relationships* (Cambridge: Cambridge University Press, 1990), 122-23.

capacity to apprehend the divine . . . is lamed in the same measure as is his capacity to experience a reality absolutely independent of himself.[67]

"The unbelieving marrow of the capricious man," Buber observed elsewhere, "cannot perceive anything but unbelief and caprice, positing ends and devising means. His world is devoid of sacrifice and grace, encounter and presence, but shot through with ends and means."[68] Buber concluded *I and Thou* by warning his readers not to try to divide their lives between what they imagined to be an "actual" relationship to God and an "inactual," or I-It relation to the world. "Whoever knows the world as something to be utilized," he wrote, "knows God in the same way."[69]

Returning to Heidegger's argument, the modern habit of enframing has revealed a "world" in which it has become possible for us to take control of our circumstances in any number of respects. Yet enframing has also concealed other ways of revealing that might, in the sense of *poiesis*, better let others (including nature) come into "presence" as truly themselves. "The threat to man does not come in the first instance from the potentially lethal machines and apparatus of technology," Heidegger observes. "The actual threat has already affected man in his essence. The rule of Enframing threatens man with the possibility that it could be denied to him to enter into a more original revealing and hence to experience the call of a more primal truth."[70]

What is this "more primal truth"? Heidegger suggests that we see it reflected in the original meaning of the Greek work *techne*. "There was a time," he writes, "when it was not technology alone that bore the name *techne*. Once that revealing that brings forth truth into the splendor of radiant appearing also was called *techne*. . . . there was a time when the bringing-forth of the true into the beautiful was called *techne*. And the *poiesis* of the fine arts also was called *techne*."[71] "In Greece," Heidegger continues,

> at the outset of the destining of the West, the arts soared to the supreme height of the revealing granted them. They brought the presence of the gods, brought

---

[67]Martin Buber, *Eclipse of God: Studies in the Relation Between Religion and Philosophy* (New York: Harper & Row, 1952), 14.

[68]Buber, *I and Thou*, 110.

[69]Ibid., 156.

[70]Heidegger, "The Question Concerning Technology," 28.

[71]Ibid., 34.

the dialogue of divine and human destinings, to radiance. And art was simply called *techne*. It was a single, manifold revealing. It was pious, *promos*, i.e., yielding to the holding-sway and the safekeeping of truth.[72]

"The arts," as Heidegger uses the term here, are not simply to be understood as one sector of cultural activity among others. Neither were they merely appreciated aesthetically. Instead, Heidegger alleges that classical Greek art arose out of and discloses an entire way of being-in-the-world, a way of being that was properly ordered by *poiesis*. He believes it still is possible to catch a glimpse of this earlier, richer way of being-in-the-world in the Greek language itself; there seems to have been an innocence, a kind of freshness, and a genuine sense of wonder literally built into words like *aletheia* (truth). *Aletheia* was not originally understood to be the product of method but was understood to be a kind of revelation awarded only to those who had paid very close attention to—and had therefore marveled at—that which is.

Heidegger goes on to contend that the Western habit of enframing, inaugurated already in the philosophies of Plato and Aristotle, occluded the original innocence and sense of wonder that had characterized pre-Socratic Greek thought—and that this, finally, is why modern Western thinkers have so often found the world to be sterile and meaningless apart from their imposition of meanings upon it. It is also, ironically, why our imposition of meanings upon the natural world fails to yield lasting satisfaction. George Grant captures this quite poignantly in his analysis of the North American *pathos*. What is now absent for us, Grant grieves,

> is the affirmation of a possible apprehension of the world beyond that as a field of objects considered as *pragmata*—an apprehension present not only in its height as "theory" but as the undergirding of our loves and friendships, of our arts and reverences, and indeed as the setting for our dealing with the objects of the human and non-human world. Perhaps we are lacking the recognition that our response to the whole should not most deeply be that of doing, nor even that of terror and anguish, but that of wondering or marveling at what is, being amazed or astonished by it, or perhaps best, in a discarded English usage, admiring it; and that such a stance, as beyond all

---

[72]Ibid.

bargains and conveniences, is the only source from which purposes may be manifest to us from our necessary calculating.[73]

Heidegger believed that what we have termed the mechanical worldview can be traced all the way back to the beginnings of what would become the Western intellectual tradition, and to a kind of neglect of a deep appreciation for *Being*. This neglect has meant, as Borgmann attempts to summarize Heidegger's thesis, that the "ground on which everything rests and the light in which everything appears moved into oblivion."[74] The neglect of Being appears to have stemmed, at least in part, from the Platonic insistence that "knowing" ought to be understood on the analogy of sight and seeing, which had (and still has) the effect of removing the one who knows—now the "observer"—from any kind of genuine existential involvement with that which is known or observed. Platonic knowers (read: observers) have thus tended to become the sole "subjects" amidst a world of mere "objects."

The bifurcation of the world into subjects and objects would take centuries to come to fruition, however, and it really only does so in the seventeenth century in the philosophy of Descartes and in the subsequent rise of early-modern science. As George Steiner comments in his work on Heidegger, "For Descartes' truth is determined and validated by certainty. Certainty, in turn, is located in the ego. The self [therefore] becomes the hub of reality and relates to the world outside itself in an exploratory, necessarily exploitative, way. As knower and user, the ego is predator."[75] "Since Roman engineering and seventeenth century rationalism," Steiner continues,

> Western technology has not been a vocation but a provocation and imperialism. Man challenges nature, he harnesses it, he compels his will on wind and water, on mountain and woodland. The results have been fantastic. Heidegger knows this: he is no Luddite innocent or pastoralist dropout. What he is emphasizing is the price paid. Things, with their intimate, collaborative affinity with creation, have been demeaned into objects.[76]

---

[73]Grant, "In Defense of North America," 35.
[74]Albert Borgmann, *Technology and the Character of Contemporary Life: A Philosophical Inquiry* (Chicago: University of Chicago Press, 1984), 39.
[75]George Steiner, *Martin Heidegger: With a New Introduction* (Chicago: University of Chicago Press, 1989), 31.
[76]Ibid.,139.

Heidegger's grand project, then, entailed tracing this original neglect of Being through the stages of thought advanced in the modern period by Descartes, Leibniz, Kant, Hegel, and—finally—Nietzsche. Indeed, Heidegger believed that Nietzsche's "will-to-power" represented the culmination of Western humanity's drive to establish itself absolutely, independent of any ground that might be said to take priority over the human will.[77] Modern technology, he believed, is simply the material manifestation of Nietzsche's metaphysical announcement of nihilism.

If modern technology obscures Being rather than bringing it to light, it thereby obscures the human essence.[78] The human relation to otherness should not, Heidegger felt, be one of grasping and pragmatic use. Rather, it ought to be a relation of audition, of privileged listening. The truly and uniquely human vocation is, in effect, to listen for the voice of Being and to give expression—primarily in language, but also in *technē*—to what we have, in this sense, heard. The genuinely human task in the world is, then, by virtue of the apprehension of otherness, to let things come into being as truly themselves—so that they can "be" in the fullest sense. Rüdiger Safranski thus notes in his intellectual biography of Heidegger:

> Whichever way one tries to tackle the problem, it ultimately remains a re-statement of Schelling's wonderful idea that nature opens its eyes in man and notices that it exists. Man as the place of self-visibility of Being. "Without man Being would be mute; it would be there, but it would not be the True one."[79]

Heidegger's provocative thesis was nicely summarized by his English-speaking editor, William Lovitt, in the introduction to the 1977 Harper & Row edition of *The Question Concerning Technology*:

> Heidegger sees every aspect of contemporary life, not only machine tech-nology and science but also art, religion, and culture understood as the pursuit of the highest goods, as exhibiting clear marks of the ruling essence of technology that holds sway in the dominion of man as self-conscious, representing subject. Everywhere is to be found the juxtaposing of subject

---

[77]Ibid.
[78]Ibid.
[79]Rüdiger Safranski, *Martin Heidegger: Between Good and Evil*, trans. Ewald Osers (Cambridge, MA: Harvard University Press, 1998), 369.

and object and the reliance on the experience and evaluating judgment of the subject as decisive. The presencing of everything that is has been cut at its roots. Men speak, significantly enough, of a "world picture" or "world view." Only in the modern age could they speak so. For the phrase "world picture" means just this: that what is, in its entirety—i.e., the real in its every aspect and element—now is "taken in such a way that it first is in being and only is in being to the extent that it is set up by man, who represents and sets forth." Were contemporary man seriously to become aware of this character of his life and of his thinking, he might, with the modern physicist, well say, "It seems as though man everywhere and always encounters only himself."[80]

Rather than existing in dialogue with the larger world that exists outside of our heads, including that even of our own bodies, Heidegger believed that we moderns—having been conditioned to *see* the world in one particular way for millennia—have instead become narrowly preoccupied with our own projects of making and with the "uses" that we can find within these projects for the "stuff" that we find in the world. We have, therefore, become very largely blind and deaf to the marvelous "otherness" of the world.

Is There Any Remedy?

We began this chapter by suggesting that the contemporary lack of concern over the drift of automatic machine technology away from ordinary embodied human existence extends beyond the impact that the technological workplace is probably having upon us, beyond the common assumption that more and newer technology is always better, and even beyond the influence of all those who stand to benefit from our opting in to more and newer technologies. Rather, our acquiescence in the face of technological disembodiment owes, ultimately—and perhaps most importantly—to a peculiar view of the world. That view sees nature, including human nature, as a vast and elaborate mechanism that differs from automatic machine technology in degree but not in kind. Embodied human existence from within this worldview is, again, most often construed, not as something to be nurtured

---

[80]William Lovitt, introduction to *The Question Concerning Technology and Other Essays*, trans. William Lovitt (New York: Harper & Row, 1977), xxxiii.

and enhanced (perhaps *by* technology) but rather as a series of limitations to be overcome with more and better technology.

As we have now seen, this mechanical world picture has deep and extensive roots within the Western tradition. To use Charles Taylor's terms, modern culture is characterized by an "instrumental stance" toward life, a stance that is "overdetermined" in the sense that it has arisen from a number of different sources and is even now buttressed by compelling convictions concerning the meaning and purpose of human life. Not only is the instrumental stance supported by modern science, Taylor observes, but it has also become, largely by way of religious convictions, central within the modern ethical outlook. That outlook continues to place a high value upon taking rational and efficacious control of all things by way of methods, procedures, techniques, and technologies.[81]

As we have also seen, the mechanical world picture underwrites the legitimacy of ongoing technological development even as it (somewhat ironically) elucidates the plausibility of soft nihilism within contemporary culture. We have managed, on the basis of this world picture, to establish a measure of control over our world. Yet now, the confidence with which the modern technological project began has seemed to vanish into postmodern nothingness. As Buber lamented,

> In sick ages it happens that the It-world, no longer irrigated and fertilized by the living currents of the You-world, severed and stagnant, becomes a gigantic swamp phantom and overpowers man. As he accommodates himself to a world of objects that no longer achieve any presence for him, he succumbs to it. Then common causality grows into an oppressive and crushing doom.[82]

This peculiarly modern problem must not be understood romantically or sentimentally. The point is not simply that we have lost a kind of natural simplicity or innocence to which we must now strive somehow to return. Rather, what has been lost, as we have stressed repeatedly, is the possibility of encountering something outside of ourselves that might direct and discipline—and thus give order to—human making and willing. In the absence

---

[81]Taylor, *Sources of the Self*, 232.
[82]Buber, *I and Thou*, 102-3.

of such discipline, modern machine technology appears destined to become more automatic, increasingly autonomous, and progressively removed from the needs and requirements of ordinary embodied human beings.

So what is to be done? The short answer is that we need a change of mind, a change of outlook. We need to discover a new way of seeing the world or, perhaps more to the point, of *attending* to it that doesn't simply "enframe" the world as stuff to be put to use. We need to discover an alternative way of being-in-the-world that enables us to know and to make in such a way that created nature—including other people and, indeed, ourselves—is enabled to become more and not less of itself by virtue of our interaction with it. We need to reimagine our place and task in the world, furthermore, such that we might grow into that truly human vocation of caring for each other, for ourselves, and for created nature; of becoming, as Heidegger once put it, the "shepherds of being."[83] Heidegger's dialectical anthropology—in which it is recognized that we become most truly ourselves only in proper relation to otherness—provides, I think, an important clue as to how we might begin to think about doing these sorts of things.

Heidegger's suggestion echoes, however distantly,[84] the importance that Christian theology ascribes to relationality in the traditional conviction that the three persons of the Holy Trinity—the Father, the Son, and the Holy Spirit—are themselves only in relation to and with each other. From a Christian theological perspective, human persons—including our bodies— have been designed for relationship: for relationship with each other, for relationship with God, and for embodied relationship in and with the rest of created nature.

We will spend much of the next chapter looking at the astonishingly high view of human embodiment that is entailed in the Christian doctrines of incarnation and resurrection. Embodied human existence is so important within God's good creation that God the Son—the One who lives and reigns with the Father and the Spirit, One God, for ever and ever—has become permanently and irrevocably an embodied human being. Christ's glorious

---

[83]See Safranski, *Martin Heidegger,* 368.
[84]The distance is admittedly great, particularly considering Heidegger's refusal ever to recant his close association with National Socialism in Germany.

resurrection from the dead, furthermore, signals that God the Father remains absolutely committed to a created order that includes human bodies. Within Christian understanding, therefore, there is no room whatsoever for any kind of gnostic denial of the body. In addition, and as we will see, taking the incarnation seriously puts an entirely new spin on what we ought to use our technologies for.

The intriguing suggestion that our task in the world ought to be "to give voice to being" offers a second clue as to how we might redress the problem of enframing. In this endeavor we hear echoes—again, however distant—of the Christian doctrine of creation, in which the human task in the world is understood in terms of naming, of stewarding, and of tending the creatures, all the while offering created order back to its Creator in words of thanks and praise. In what follows, we will seek to remember from the perspective of the Christian faith just what kind of world has been given to us to live in and just what kind of work we have been given to do in this world. Remembering both things will be crucial if we are to redirect modern technological development back toward the needs and requirements of ordinary embodied human persons.

CHAPTER FOUR

REMEMBERING WHERE WE ARE
AND WHO WE ARE

We should rejoice and be in wonder that our Lord Jesus Christ was
made man, rather than that he, as God, performed divine deeds
among men. Our Salvation, after all, depends more upon what
he was made on our behalf, than on what he did among us.

ST. AUGUSTINE, "HOMILY 17,"
HOMILIES ON THE GOSPEL OF JOHN 1-40

CHRISTIANLY SPEAKING, the observation that automatic machine development is deviating away from ordinary embodied human existence needs to be followed by the proclamation of the embodied and deeply humanizing implications of the gospel of Jesus Christ. Such a proclamation is the gist of this fourth chapter. There are, however, several features of contemporary culture—as well as of contemporary Christian culture—that need to be appreciated before we can make this announcement. For not only has contemporary secular culture fallen asleep to the dangers of technological dehumanization, but many of our churches also appear to have forgotten how central ordinary embodied human existence is to Christian faith.

That modern secular culture has not offered much in the way of resistance to the dehumanizing thrust of automatic machine development is unfortunate, though perhaps understandable. Given the material fruits of technological development hitherto, given its alluring promises of "more," "better," "faster," and "easier," given the powerful monetary interests that attach to creative/destructive innovation, and given that the modern mind remains—Romantic protests notwithstanding—by and large mechanical in its outlook, it is not terribly surprising that modern secular thought has been largely acquiescent in the face of recent technological developments. While this could change, it is not clear that modern secular thought possesses the intellectual wherewithal to mount much of a protest. Protests thus far have been raised in the name of rights, either to privacy or on behalf of groups believed to be deserving of a larger share of technology's material benefits.[1] Still, they have not as a rule taken issue with the drift of automated machinery away from ordinary embodied human existence. While authors like Nicholas Carr have sought to sound the alarm over this drift, the postmodern intellectual environment has made it almost impossible to protest this development in the name of truth. Yet the momentum and inertia of modern technological development are not likely to be alterable in the name of anything less.

It is genuinely tragic, then, that our churches have not offered more considered resistance to modern automatic machine development, for the extraordinarily high value of ordinary embodied human being is a core Christian truth. Yes, the Christian hopes for salvation *from* "this world" and for an "eternal life" in a "world to come," but this hope has never—within the broad tradition of Christian orthodoxy—been understood in such a way as to disparage ordinary embodied human existence. Neither has the church ever before kept silent in the face of gnostic promises such as those of transhumanism. The church has long recognized that if the eternal Word of God "became flesh and made his dwelling among us," as the apostle John declares (Jn 1:14), this confers staggering value upon ordinary fleshly existence.

Modern technological development's apparent trend away from ordinary embodied human existence should have triggered alarms in our churches. The fact that it has not may signal that our churches are suffering from a kind of latent Platonism, as if being saved from what the apostles termed "this

---

[1]See, for example, David Lyon, ed., *Surveillance as Social Sorting: Privacy, Risk, and Digital Discrimination* (New York: Routledge, 2003); also Karen Mossberger, Caroline J. Tolbert, and William W. Franko, *Digital Cities: The Internet and the Geography of Opportunity* (Oxford: Oxford University Press, 2012).

world" (the *saeculum*) meant being saved from bodily existence altogether. Although consistently censured within orthodox Christianity, similar beliefs have surfaced regularly in Christian history.

In the North American context, the church's silence as modern technology has diminished embodied human being may also stem from the fact that evangelical Protestants have always—from Gutenberg's printing press to radio and television—been early adopters of modern technologies and have used them to great benefit.[2] Perhaps evangelicals assume that whatever modern entrepreneurs and engineers manage to come up with, they will find a way to render it serviceable in spreading the Christian message. It may also be that Christian acquiescence to automatic machine development is a kind of combination of mistaken "otherworldliness" and this fascination with technological possibilities.

Unfortunately, what also appears to account for the church's odd reticence in the face of the dehumanizing trajectory of modern automatic machine technology is simple forgetfulness. Our churches seem to have forgotten the centrality of embodied human existence within the gospel of Jesus Christ. Indeed, what appears to account for the church's failure to protest both the mechanical modern outlook as well as the modern tendency to objectify and enframe the natural world is that we have fallen out of the habit of reflecting upon—and living out of—the implications of core Christian convictions.

Of course, in this forgetfulness we Christians are not alone. We appear to have succumbed, along with most other traditional religious communities, to that peculiar form of amnesia that is fostered by modernization. This kind of obliviousness helps explain the marginalization of traditional religion within increasingly secular modern societies as well as the odd subjectivity and eccentricity of religious belief even within what were once traditional religious communities.[3] As Daniéle Hervieu-Léger has observed, "The growth of secularization and the loss of total memory in societies without a

---

[2]See Quentin J. Schultze, *Christianity and the Mass Media in America: Toward a Democratic Accommodation* (East Lansing: Michigan State University Press, 2003).
[3]See, for example, Ross Douthat, *Bad Religion: How We Became a Nation of Heretics* (New York: Free Press, 2012).

history and without a past coincide completely; the dislocation of the structures of religion's plausibility in the modern world works in parallel with the
advance of rationalization and successive stages in the crumbling of collective memory."[4] Contrary to received secular opinion, Hervieu-Léger
contends that modern societies have neither outgrown nor found secular
substitutes for religion, and neither are they more rational than traditional
societies. Rather, modern societies have simply become "amnesiac," in some
cases repudiating and in others simply failing to maintain the "chain of
memory" that binds them to their religious pasts. For modern Christian
churches, this has meant forgetting (among other things) the shape and
implications of basic Christian theology. Having forgotten this theology, it
has become all but impossible for the church to speak prophetically into
modern society and culture about such things as the dehumanizing trajectory of modern technological development.

The Christian response to what we have called the disembodying or *dis*-
membering arc of automatic machine technology must, therefore and in the
first instance, be one of *re*-membering. We will need to remember why, from
the point of view of the Christian religion, disembodiment might be a
problem. We will also need to remember why the mechanical world picture
as well as the modern habit of enframing must be very largely rejected and
supplanted with an outlook—and practices—that celebrate God's good creation and our particular vocation within the created order.

This fourth chapter, then, is about remembering. It is about remembering
what the Christian religion says about *where* we are and that we live in a world
that we did not make. It is about remembering *who* we are and that it is not
entirely up to us to decide who we will be in this world that we did not make.
And it is about remembering how, from the point of view of the Christian
religion, we ought to understand the human task—our *vocation*—within this
world that we did not make. The point of remembering these remarkable
things in the present context is so that we might begin again to think clearly
about modern technological development. In this light we might hopefully
begin to do what we can to redirect modern technology toward enabling

---

[4]Daniéle Hervieu-Léger, *Religion as a Chain of Memory* (New Brunswick, NJ: Rutgers University
Press, 2000), 127.

created things, including those creatures that were created after the "image" and "likeness" of God, to become more and not less of themselves.

It may come as a surprise to the contemporary Christian reader that some of the thinking that currently informs modern technological development is almost entirely at odds with biblical religion. Basically, neither the mechanical world picture nor the objectification that such a view of the world appears to foster are at all compatible with biblical understanding. This means that renewed Christian reflection about modern technology must, in the first instance—and for all of our constructive intentions—enter our technological culture as criticism, and we should not expect it to be welcomed. Yet, and as we said at the beginning of the chapter, our concerns about the incorporeal and dehumanizing arc of modern technological development must very quickly give way to a proclamation of the embodied and deeply humanizing prospects of the gospel of Jesus Christ. The following theological recollection, then, is organized around the basic story elements of the biblical narrative: creation, fall, redemption, and consummation. While we will only be able to introduce these elements, the point of our recollection is simply to make it clear that each has tremendously significant implications for assessing modern technological development going forward. Taken together, they help us to understand where we are, who we are, and the kind of work we have been called—often by way of inventive technologies—to do.

Creation

The Christian faith affirms that, far from being the product of blind chance and impersonal mechanical forces, the world actually owes its existence to God, who, as the Nicene Creed states, created "all things visible and invisible." The world is, furthermore, a work of love. The creation bespeaks the care and loving-kindness of its Creator. It is not comprised of meaningless stuff, and it is far less like a machine and/or mechanism than it resembles a lovingly crafted work of art. Even as the work of human artists calls for careful consideration and often admiration, so the Christian religion asserts that we are to marvel at, be amazed by, and be deeply grateful for God's marvelous creation of all things. Yet whereas human artworks are crafted in two, three, or possibly four dimensions, God's creation is enormously more nuanced, including even the free agency of his creatures in its design—in ways that extend beyond human reckoning. "Where were you when I laid the earth's foundation?" God questions Job. "Tell me, if you understand. Who marked off its dimensions? Surely you know!" (Job 38:4-5). But of course, Job does not know. Eventually he confesses that the creation bespeaks things that are simply "too wonderful" for him to know (Job 42:3).

Christian understanding of creation as the work of love arose out of the revelation in Jesus Christ that God is triune, which is to say that God is not ultimately a monolithic and irresistible force, but rather that he is manifold and radically personal. God has existed from eternity in the interpersonal relatedness of the Father, the Son, and the Holy Spirit. The implications of God's tri-unity spread out in all directions. It means, for example, that the creation was (and is) not necessary. The creation cannot be understood as a kind of emanation of "divinity," over which God must somehow preside in order to be God. And neither can creation be understood as the kind of thing that must at some point inevitably collapse back into the unity of the godhead. Rather, the Christian religion asserts that God created all things "out of nothing" (*ex nihilo*) simply—and wonderfully—so that they might glorify him by participating in the love that has from eternity animated the "being in communion"[5] that was, that is, and that ever will be "the Father, the Son, and the Holy Spirit."

The revelation of God as Trinity means that everything that "is" is the product of personal intention, that all things exist by virtue of the original and ongoing determination of the One who brought them into and who continues to hold them in existence. Yet while the creation clearly reflects God's sovereign will, this must not be understood simply in terms of the will-to-power. For again, although the Christian religion does assert that God is all-powerful (*omnipotent*), so it further affirms that God's power is always in the service of his love, and hence that the creation of all things is above all an expression of God's will-to-love. Again, created things were (and are) brought into existence in order, finally, to glorify God by participating in his love.

---

[5] I am borrowing this phrasing from John Zizioulas, *Being as Communion: Studies in Personhood and the Church* (Crestwood, NY: St. Vladimir's Seminary Press, 1985).

In respect to this participation, Christian theologians have from the beginning insisted that created things enjoy real and not simply apparent existence, for love requires the gift of oneself to another. While no-thing exists independently of God, all things are nevertheless "other" than God. This "otherness" is itself the gift of God. It is what enables created things to "be" and to become the recipients of his love. The very mystery of human and angelic freedom—which is to say, of *spiritual* freedom—appears to originate in this gift of genuine otherness.

Just as the Christian religion affirms that the Father, the Son, and the Holy Spirit are who they are only in relation to one another, so all of the things that have been willed into existence by the triune God exist primarily in relation. They exist in relation to God, their Creator, and yet they also exist in relation to one another. Indeed, reflecting the nature of their triune and interpersonal Creator, created things become what and who they are only in their relations to other things. This is most obviously true of the animals and of God's human creatures, whose relational nature is emphasized in the Genesis narrative in the refrain that God created them "male and female" (Gen 1:27). Obedience to the first of the divine commands, "Be fruitful and increase in number" (Gen 1:28), thus required the intimate relation of two fundamentally different and yet clearly complementary aspects of being. In the case of the human creatures, once again reflecting the fact that while "the Father, the Son, and the Holy Spirit" are three persons they are nevertheless one God, the original pair are said to have become "one flesh" in the intimacy of sexual union. That out of this union of the man and the woman new life should arise has long been seen to epitomize the fecundity of love, as well as to reflect (however dimly) the trinitarian relations of Father, Son, and Holy Spirit.

The Christian doctrine of creation further affirms that the creation is good. It is not, as has long been asserted by gnostics, a problem that must somehow be overcome. The goodness of creation is manifest in its astonishing diversity. Indeed, far from being a defect, the diversity and pluriformity so evident in creation reflect divine delight and beneficence as well as the infinite extent of the divine nature. As no single creature could possibly reflect the entirety of his goodness, so God created a great many creatures, each reflecting his nature and character in some small, perhaps, yet never-theless unrepeatable fashion. Again, the Christian doctrine of creation stresses that each and every individual creature exists because God loves it and desires that it should exist. The fourteenth-century Christian mystic Julian of Norwich captured this in the following justly celebrated passage:

> The Lord showed me something small, no bigger than a hazelnut, lying in the palm of my hand, as it seemed to me, and it was round as a ball. I looked at it with the eyes of my understanding and thought: What can this be? I was amazed that it could last, for I thought that because of its littleness it would suddenly have fallen into nothing. And I was answered in my understanding: It lasts and always will, because God loves it; and thus everything has being through the love of God. In this little thing I saw three properties. The first is that God made it; the second is that God loves it; the third is that God pre-serves it. But what did I see in it? It is that God is the Creator and the protector and the lover.[6]

That God made, protects, and loves each and every individual created thing is especially true of human beings, those creatures made after God's own "image" and "likeness." "Are not five sparrows sold for two pennies?" Jesus reassured an anxious crowd. "Yet not one of them is forgotten by God. Indeed, the very hairs of your head are all numbered. Don't be afraid; you are worth more than many sparrows" (Lk 12:6-7). While we tend to fit indi-viduals beyond family and friends into groups in order to consider them, often categorizing them according to race, class, gender, ability, and so forth, God knows and cares for every one of us individually. He knows each of us by name. Indeed, from the Christian point of view, as philosopher Nicholas Berdyaev once noted, "every single human soul has more meaning and value than the whole of history with its empires, its wars and revolutions, its blos-soming and fading civilizations."[7]

Christian ecclesiology further stresses the irreplaceable importance of each member of Christ's body, the church. "Now if the foot should say, 'Be-cause I am not a hand, I do not belong to the body,' it would not for that

---

[6]Julian of Norwich, cited in Ian McFarland, *From Nothing: A Theology of Creation* (Louisville: Westminster John Knox, 2014), 60.

[7]Nicholas Berdyaev, *The Fate of Man in the Modern World* (Ann Arbor: University of Michigan Press, 1935), 12.

reason stop being part of the body," the apostle Paul wrote to the Corinthian church. On the contrary, he insisted: "there are many parts, but one body" (1 Cor 12:15-26). Elaborating on the apostolic teaching, theologian Ian Mc-Farland writes:

> Difference is central to life before God because God calls each human being to a different place within the body of Christ. Human beings are therefore equal in that they are all called by God to be persons in Christ (so that their equality is grounded extrinsically in God rather than in any intrinsic attribute or property they possess); but they differ in that they are called to enact that personhood in distinct and unsubstitutable ways. All are called to live under Christ, the one head, but no two people occupy the same place in the body.[8]

Christianly understood, the ground of human dignity as well as the highest human calling are tied up in individual, responsible, and personal agency before God and neighbor. Even secular Western concerns and priorities continue to reflect—however dimly and occasionally ironically—these originally Christian theological convictions. To be deemed a "responsible" person still is high praise, and to be "held responsible" remains a serious charge. For this reason, as philosopher William Barrett once observed, even modern secular Western people continue, in large part, to be de facto Judeo-Christians.[9]

Beyond its diversity and pluriformity, the deep goodness of creation is evident in its having been perfectly "fit" to the purpose of glorifying its creator, which it does by living, indeed by thriving. God's good creation has been ordered to fullness, to diversity, to pluriformity, to multiplicity, to beauty, to fruition, and to growth. Theologian Jonathan Wilson employs the term "superabundance" in an attempt to capture the particular way in which the creation is good.[10] In a little essay entitled "Let Me Tell You Why," theologian Robert Farrar Capon playfully expresses the divine delight that lies behind the creation's superabundance:

---

[8]McFarland, *From Nothing*, ix.

[9]William Barrett, cited in William Poteat, *Polanyian Meditations: In Search of a Post-Critical Logic* (Durham, NC: Duke University Press, 1985), 127.

[10]Jonathan R. Wilson, *God's Good World: Reclaiming the Doctrine of Creation* (Grand Rapids: Baker Academic, 2013), 18.

One afternoon, before anything was made, God the Father, God the Son, and God the Holy Spirit sat around in the unity of their Godhead discussing one of the Father's fixations. From all eternity, it seems, he had had this *thing* about being. He would keep thinking up all kinds of unnecessary things—new ways of being and new kinds of beings to be. And as they talked, God the Son suddenly said, "Really, this is absolutely great stuff. Why don't I go out and mix up a batch?" And God the Holy Spirit said, "Terrific! I'll help you." So they all pitched in, and after supper that night, the Son and the Holy Spirit put on this tremendous show of being for the Father. It was full of water and light and frogs; pinecones kept dropping all over the place and crazy fish swam around in the wine glasses. There were mushrooms and mastodons, grapes and geese, tornadoes and tigers—and men and women everywhere to taste them, to juggle them, to join them, and to love them. And God the Father looked at the whole wild party and said, "Wonderful! Just what I had in mind! *Tov*! [good!] *Tov*! *Tov*!" and all God the Son and God the Holy Spirit could think of to say was the same thing, "*Tov*! *Tov*! *Tov*!" So they shouted together "*Tov meod*!" And they laughed for ages and ages, saying things like how great it was for beings to be and how clever of the Father to think of the idea, and how kind of the Son to go to all that trouble putting it together, and how considerate of the Spirit to spend so much time directing and choreographing. And for ever and ever they told old jokes, and the Father and the Son drank their wine in *unitate Spiritus Sancti* [the unity of the Holy Spirit], and threw ripe olives and pickled mushrooms at each other *per omnia saecula saeculorum* [forever and ever]. *Amen*.[11]

Capon is quick to acknowledge that his is a crass analogy. Still, he reasons that such are perhaps best when trying to capture the essence of something, the fullness of which must certainly lie beyond human comprehension. In the case of creation, Capon suggests in classically Christian fashion, this essence arises out of the love, the fellowship, and the sheer joy that have been shared by the Father, the Son, and the Holy Spirit, world without end.

The creation has thus been ordered to fruition. It has been wonderfully and carefully crafted by God to make life possible, which is why Christian theology has traditionally spoken, not just of creation, but also of created order. Comprised of matter and movement ordered according to what today

---

[11]Robert Farrar Capon, "Let Me Tell You Why," in *The Romance of the Word: One Man's Love Affair with Theology* (Grand Rapids: Eerdmans, 1996), 176-77.

we understand in terms of physical laws and constants, the notion of created order delineates a framework—physical, moral, and spiritual—within which the varied and sundry creatures are enabled to live and move and—hopefully—thrive. This is why the Genesis narrative describes the creation of spaces before detailing the several categories of creatures that will live in them, in effect describing habitats ordered to the life and flourishing of inhabitants. The separation of light from darkness, of the waters above from the waters below, of the sea from the dry land—all of these separations prepare spaces where things can happen and the various kinds of creatures can exist: the fish in the seas, the animals on the dry land, the birds in the air, and for the creatures created after the image and likeness of God, the entire earth. As O'Donovan notes,

> Creation is the given totality of order which forms the presupposition of historical existence. "Created order" is that which is not negotiable within the course of history, that which neither the terrors of chance nor the ingenuity of art can overthrow. It defines the scope of our freedom and the limits of our fears.[12]

In one important sense, then, Christian theology affirms that the creation is complete.[13] It has been ordered, and is even now held in order by God such as to make creaturely life, including human history, possible.

As O'Donovan's comments suggest, furthermore, our own being and thriving takes place in time, which, together with space, is a basic condition of all creaturely existence. While space may well describe the where of our bodies, only time will tell just who we are and who and/or what we are to become. Responsible personal agency, in short, is historical; it is circumscribed by and unfolds within time. In this connection, Christian theology has stressed that all creaturely existence, being finite and in itself incomplete, moves inexorably toward the perfection that is God, and that this movement toward God takes time. Here again, this is not a defect, but it simply reflects God's manifest desire that everything should happen in order and that it should not all happen at once.[14] "Like a piece of music," theologian Colin Gunton noted,

---

[12]Oliver O'Donovan, *Resurrection and Moral Order: An Outline for Evangelical Ethics* (Grand Rapids: Eerdmans, 1986), 60.

[13]Ibid.

[14]Colin E. Gunton, *The One, the Three and the Many: God, Creation and the Culture of Modernity*

[the creation's] peculiar perfection consists in the fact that it takes time to be what it is. *In that respect* it is not ontologically inferior to that which is eternal, but merely different. In the goodness of God who created it as it is and directs it to its end, it just is like that.[15]

The importance of time within the created order is emphasized in the Genesis narrative by the reprise of mornings and evenings and in the succession of the days of creation. In biblical understanding there is, as the writer of Ecclesiastes memorably put it, "a time for everything, and a season for every activity under the heavens" (Eccles 3:1).

Space and time together thus condition the possibility of and possibilities for creaturely existence. Together they comprise the stage upon which the drama of life can unfold and be lived. Indeed, space and time combine to make possible the places where things can actually happen. A place, in this sense, has a location, but it also has a past, a present, and a future. Not only can things happen in a place, but things have happened and perhaps will happen there. Place may thus be said to be a basic condition for the possibility of created things to live and to be present to one another.[16] For creatures to affect and to be affected by other creatures, they must share a place.

In this connection, it is important to stress that personal agency inevitably takes place in and necessarily only makes sense in terms of particular places. For, again, not only does a place describe the where of my particular body, but it delineates the possibilities of where, how, when, and with whom I can move into the future. A place, furthermore, is where it is possible for me to feel at home. In fact, a colloquial synonym for home is "my place." As philosopher William Poteat observed, "A place is full of objects and of relations upon which I have left my very personal stamp, expressed my own idiosyncrasy, part of my unique history. It is a domicile for the love that issues from the very center of my person."[17]

This is perhaps why it is so disorienting to be deprived of our places, as when, for example, the homes we grew up in are torn down and redeveloped.

(Cambridge: Cambridge University Press, 1993), 94.

[15]Ibid., 81 (emphasis in original).

[16]McFarland, *From Nothing*, 65.

[17]William H. Poteat, *The Primacy of Persons and the Language of Culture: Essays by William H. Poteat*, ed. James M. Nickell and James W. Stines (Columbia: University of Missouri Press, 1993), 34.

We feel the impact of this loss even if we had not lived there for many years; we survive the loss, for the most part, because we have long since found other places that we know and in which we are known. Were we to have no place, we would be lost. Indeed, to have no place, Poteat observes in this connection, is to lack the very ground of being and continuing to become a human person. To be deprived of place, in short, would be to lack the minimum conditions for remaining a human person.[18]

It is hardly surprising that the provision of place is a crucial biblical theme. Indeed, the ordering of places and of times is regularly and repeatedly celebrated in the Scriptures and ascribed to God's sovereignty, goodness, loving-kindness, and faithfulness, as the psalmist declares in Psalm 33:4-9

> The word of the LORD is right and true;
>   he is faithful in all he does.
> The LORD loves righteousness and justice;
>   the earth is full of his unfailing love.
> By the word of the LORD the heavens were made,
>   their starry host by the breath of his mouth.
> He gathers the waters of the sea into jars;
>   he puts the deep into storehouses.
> Let all the earth fear the LORD;
>   let all the people of the world revere him.
> For he spoke, and it came to be;
>   he commanded, and it stood firm.

We stress the importance of *time* and *place* for embodied personal existence here, not simply because both tend to be neglected aspects of the creation narrative, but more importantly because both have come under assault in our progressively abstract and increasingly fast-paced technological society. We will return to this point in our final chapter.

It is also important to stress that when we say that the creation is "complete," this should not be taken to mean that there is nothing left to happen. The Genesis narrative implies that in the beginning the creation, while ordered to fruition, was nevertheless wild and needed still to be tamed and stewarded in order to reach its full potential. It remained for the last of those to be created,

---

[18]Ibid.

for the human creatures, to carry out this important task. Thankfully, they were specially equipped to do this work—for unlike any of the other creatures, they were created after the image and likeness of God (Gen 1:26).

Just what the "image of God" (*imago Dei*) means has long been debated within the Christian tradition. Scientist and educator Leon Kass notes that up to this point in the Genesis narrative, a number of important things have already been revealed about the nature and character of God: that he speaks, commands, names, blesses, and hallows; that he makes and makes freely; that he looks at, beholds, and delights in what he has made; that he graciously provides for his creatures; and that he anticipates the completion of his creation.[19] To bear the image and likeness of such a God, then, would appear to suggest being able to do many of these same kinds of things. "Human beings," Kass continues,

> alone among the creatures, can articulate a future goal and use that articulation to guide them in bringing it into being by their own purposive conduct. Human beings, alone among the creatures, can think about the whole, marvel at its many-splendored forms and articulated order, wonder about its beginning, and feel awe in beholding its grandeur and in pondering the mystery of its source.[20]

Indeed, the very fact that we are able to read and ponder the meaning of the Genesis narrative is perhaps itself a clear indication of the specific shape of the uniquely human place within the created order. "Reading Genesis 1," Kass writes, "performatively demonstrates the truth of its claims about the superior ontological standing of the human."[21] The creatures bearing the divine image are made, the text tells us, "so that they may rule over the fish in the sea and the birds in the sky, over the livestock and all the wild animals, and over all the creatures that move along the ground" (Gen 1:26).

Just as the meaning of the *imago Dei* has been debated within the Christian tradition, so too has the notion of "ruling over" been subject to a variety of interpretations. The consensus today is that the parallel creation accounts in Genesis 1 and 2 describe the creation of the world on the analogy of the

---

[19]Leon R. Kass, *The Beginning of Wisdom: Reading Genesis* (New York: Free Press, 2003), 37-38.
[20]Ibid., 38.
[21]Ibid.

construction and dedication of an ancient temple, a sacred space in which the creatures who bear the image and likeness of God have been placed to represent God by mediating his essence (love) and intention (fruition) both *to* and *for the sake of* all of the other creatures that have also been placed within the sacred space.[22] In so representing the divine essence and intention within the creation temple, the human stewards prepare that place where God himself will ultimately dwell. As O'Donovan observed, the human rule over the rest of creation is—or should be—the rule that "liberates other beings to be, to be in themselves, to be for others, and to be for God."[23]

This liberation of the creatures to be and to become most fully themselves is to be accomplished, at least in part, by means of speech. For human beings, alone among God's creatures, are permitted to name the other creatures and, in effect, to return the creation to its creator in the speech of thanksgiving and praise. The human liberation of created nature is also accomplished by means of creative human labor. As Romanian Orthodox theologian Dumitru Staniloae once observed,

> The gifts given to us by God can become our gifts to God through the fact that we are free to give things back to God. We transform things into our gifts by the exercise of our freedom and by the love which we show to God. Towards this end we are able to transform and combine them endlessly. God has given the world to man not only as a gift of continuous fruitfulness, but as one immensely rich in possible alterations, actualized by each person through freedom and labour. This actualization, like the multiplication of talents given by God, is the gift of humankind to God.[24]

Along with all of the other gifts God has given us, and as astonishing as it may sound, it has also been given to us in Christ to give the world back to God in works of love. The world is not simply objectified when used in this

---

[22]See, for example, John H. Walton, *The Lost World of Genesis One: Ancient Cosmology and the Origins Debate* (Downers Grove, IL: InterVarsity Press, 2009); also Rikk E. Watts, "On the Edge of the Millennium: Making Sense of Genesis 1," in *Living in the Lamblight*, ed. Hans Boersma (Vancouver: Regent College, 2001), 129-51; also Rikk E. Watts, "The New Exodus/New Creational Restoration of the Image of God," in *What Does It Mean to Be Saved?*, ed. John G. Stackhouse Jr. (Grand Rapids: Baker, 2002), 15-41.

[23]O'Donovan, *Resurrection and Moral Order*, 38.

[24]Staniloae, cited in Charles Miller, *The Gift of the World: An Introduction to the Theology of Dumitru Staniloae* (Edinburgh: T&T Clark, 2000), 62.

way but is mysteriously enabled to achieve *presence* in a new and profound way. Thus, while the things we find in the world do not become valuable simply because we desire them enough to place a value on them, perhaps paying money for them, they are strangely—and wonderfully—magnified when used in the service of love.

When it comes to remembering where we are, then, the Christian religion affirms that we are creatures who have, along with a great many other creatures, been made to inhabit God's good creation, something for which we should be deeply grateful. As Psalm 100 declares,

> Shout for joy to the LORD, all the earth.
>> Worship the LORD with gladness;
>> come before him with joyful songs.
> Know that the LORD is God.
>> It is he who made us, and we are his;
>> we are his people, the sheep of his pasture.
> Enter his gates with thanksgiving
>> and his courts with praise;
>> give thanks to him and praise his name.
> For the LORD is good and his love endures forever;
>> his faithfulness continues through all generations.

We must also remember who we are. Christianly understood, we are the creatures who bear the divine image and who have thus, as Psalm 8 announces, been "crowned with glory and honor" to have dominion over "all flocks and herds, and the animals of the wild, the birds in the sky, and the fish in the sea, all that swim the paths of the seas" (Ps 8:7-8). It has been given to us to mediate God's love to his creation for the sake of human flourishing as well as for the sake of the flourishing of all the other creatures that have been placed within God's "Creation temple."

Fall

The modern outlook has tended to be optimistic about the human prospect, by and large envisioning the world as a machine and the human purpose in terms of design and engineering. Although grave misgivings about the modern project have come to haunt the postmodern imagination, our culture remains confident that, given the correct education and the appropriate development of technology, we will eventually manage to take control of our circumstances.

The Christian religion, of course, harbors no such illusions. While profoundly hopeful, biblical religion places very little hope in human effort in and of itself. As described in the somber third chapter of Genesis, the human situation has, almost from the beginning, been marred by the disastrous fall of human beings into sin and death. Represented in the eating of fruit taken from the forbidden "tree of the knowledge of good and evil," the original human pair succumbed to the temptation to seek autonomy and independence from God, for which they were cast out of the garden, and from whence they became subject to the futility, frustration, violence, and death that have ever since described the human condition. The larger creation, for its part, was bereft of its stewards, and so also became subject to fruitlessness and futility, symbolized in the text by the curse of the ground and by the spate of "thorns and thistles" that would ever afterward trouble human labor.

Why did this happen? How could it have happened in a creation that had so recently been declared very good? Why did a good and all-powerful God allow it to happen? Christian theologians have wrestled with these questions for a very long time, and space does not permit elaborating on the answers that have been tendered. Suffice it here simply to say, following Augustine, that God evidently judged it better to bring good out of evil than not to allow evil to exist.[25]

We might also say, following Irenaeus, that God has evidently determined that the love that human beings are to have for him, which is the human *telos* as well as the source of all wisdom, should develop over time and with experience. This is why human love for God appears not to have emerged fully developed *ex nihilo*, as it were. Rather, it was intended to grow and mature.

---

[25]Augustine, *Enchiridion* 8. In "On Rebuke and Grace," Augustine comments similarly: "Therefore we properly believe and confess to our salvation that the God and Lord of all, who created all things exceedingly good, both foreknew that evils would arise from these goods and knew that it was more appropriate for his sovereign goodness to deal well with these evils than to prevent their occurrence. He so arranged the life of angels and humans that he showed first what their free choice could achieve and then what the favor of his grace and the judgment of his justice could accomplish" (cited in J. Patout Burns, *Theological Anthropology*, ed. William G. Rusch, Sources of Early Christian Thought [Philadelphia: Fortress, 1981], 97-98).

It was for this reason that Irenaeus described the creation as "imperfect," for while it was indeed good, it had still to achieve its perfection in Christ. Yet our original ancestors evidently grew impatient. Rather than obediently growing up into the likeness of God, and so becoming capable of mediating the divine essence and intention to the rest of creation, the original human pair foolishly sought to seize godlikeness. Having given in to this original temptation, human progress was arrested and human development turned back in on itself in such a way as to become fruitless and ultimately lifeless, condemning the rest of creation similarly to futility and frustration.

Yet however the fall into sin is explained, sinful hubris has now infected every aspect of human existence. The human capacity for loving God and neighbor has, in effect, been twisted inward (*incurvatus in se*) to become the inordinate love of self, giving rise to dissipation, debauchery, shame, confusion, dissention, conflict, violence, cruelty, and absurdity. We have obstinately sought autonomy within humanly constructed "worlds" at enmity with God (Jn 15:18-19); we have succumbed to what the apostle Paul termed "the desires of the flesh" (Gal 5:16) and so have fallen victim to the havoc and fruitlessness of sensuality; we have become deceitful, deceiving ourselves as well as others; and we have become vulnerable to being deceived by all manner of false promises and false hopes. There is little need to elaborate on these things. A glance at any daily newspaper should suffice to corroborate that something has gone very wrong with the human species. We neither love God nor one another as we know we ought, and we are despoiling the natural world.

The Genesis narrative describes the original act of disobedience as having been followed by a trial at which, and following careful inquiry, discoveries are made, guilt is pronounced, and punishments are meted out. The human creatures are punished with the prospects of pain and suffering, with difficult and burdensome labor, with frustration and, finally—symbolized in the loss of access to "the tree of life"—with powerlessness in the face of biological death. The creatures who had been made "a little lower than the angels" and commissioned to rule created nature were left, it seems, to lapse back into a merely natural existence. A remedy for all of these maladies is graciously promised, but it is said to lie in the future. The human creatures must wait for it to be revealed.

The Genesis narrative goes on to depict the devastating personal and social consequences of human sin. In the text's treatment of the line of Cain (Gen 4:17-24), we read that our ancestors immediately sought to evade the curses that had been pronounced upon the ground and upon human labor by banding together in defiance of the command to "fill the earth" as well as apparently by way of employing ingenuity and technical skill. They even sought to overcome death in building the great tower on the Plains of Shinar. Yet the text indicates that these early technological feats were themselves dogged by violence, oppression, and, so it is to be assumed, human suffering. God confused the language of these early technologists, so the text suggests, for their own sakes. The assessment that, if allowed to continue their collective effort, "nothing they plan to do will be impossible for them" (Gen 11:6) is meant, it seems, to make the reader shudder. The Genesis account should not be taken to mean that God frowns upon human progress but rather that he abhors "progress" that comes at the expense of oppression and injustice and that points, therefore, only in the direction of death. In addition, while the narrative does not condemn human ingenuity, neither does it suggest that we can, by way of technical ingenuity, overcome the curses originally pronounced upon human labor following the original act of disobedience. Indeed, just as the origin of evil is ultimately inscrutable, so the text suggests that evil can only ultimately be overcome by God, for whom all things are possible. Just how and when God chooses to overcome evil, furthermore, remains for him to determine.

Redemption

The relatively terse telling of humanity's fall into sin and death introduces a much longer narrative that details humanity's redemption, ultimately, in Christ Jesus. Reading on, we learn that the purpose of the curses that follow upon the original act of disobedience, and indeed even the purpose of the sentence of death pronounced upon all men and women, is not finally punitive. Rather, our afflictions, ordeals, hardships, and ever-vexing limitations are meant to move us to seek after God and so eventually to be comforted, healed, and restored. Indeed, more than simply restored, for the good news of the Christian message is that our eventual restoration in Christ will far exceed the blessedness of our original condition. This is the meaning of the traditional churchly notion of *felix culpa*, which in the Catholic tradition is translated "happy fault," as in the Paschal Vigil Mass: *Exsultet O felix culpa quae talem et tantum meruit habere redemptorem*, "O happy fault that earned for us so great, so glorious a Redeemer." From the point of view of the Christian faith, in other words, our fall into sin and death is not simply the problem for which the coming of the Christ is the solution. Rather, our fall has become the occasion for God, as the apostle writes, to do "immeasurably more than all we ask or imagine" (Eph 3:20). That the one in whom dwells the fullness of God, the one in and through and for whom *all* things were created, the one through whom *all* things on heaven and earth shall be reconciled to God (Col 1:15-19)—that he should have become one with us, an embodied human being, is beyond astonishing!

"So many are the Saviour's achievements that follow from His Incarnation," St. Athanasius wrote in the fourth century, "that to try to number them is like gazing at the open sea and trying to count the waves."[26] Let us, however, at least briefly consider the following three: the first is the nature of God as Trinity; the second is God's ultimate intention vis-à-vis his damaged creation; and the third is a glimpse of human being-in-the-world as originally intended.

The implications of Christian understanding of God as Father, Son, and Holy Spirit spread out in all directions. As we have seen, the Christian doctrine of the Trinity suggests that the ground and grammar of all created things is deeply and indissolubly relational and personal: not only is the triune God lovely and loving, but his very being-in-communion is love. The order that God has so lovingly created and all the creatures he has created to fill this order, then, reflect his relational essence. Again, all created things are what and who they are in relation to other things. The destiny of all created things is to glorify God by sharing in the love that has, from all eternity, characterized the relations of the Father, the Son, and the Holy Spirit.

Concerning God's intention vis-à-vis a creation marred by human sin, it might once have been possible, before the coming of Christ—and particularly before his astonishing resurrection from the dead—for someone to wonder,

---

[26]Athanasius, *On the Incarnation: The Treatise De Incarnatione Verbi Dei*, trans. and ed. a religious of C.S.M.V. (Crestwood, NY: St. Vladimir's Seminary Press), 93.

as O'Donovan put the question, whether or not creation was basically a lost cause:

> If the creature consistently acted to uncreate itself, and with itself to uncreate the rest of creation, did this not mean that God's handiwork was flawed beyond hope of repair? It might have been possible before Christ rose from the dead to answer in good faith, Yes. Before God raised Jesus from the dead, the hope that we call "gnostic," the hope for redemption *from* creation rather than for the redemption *of* creation might have appeared to be the only possible hope.[27]

We will return to this point below, for it would appear that not a few apologists for modern technological development promise not simply to relieve us of certain painful aspects of our present condition but to free us entirely from what they perceive to be the constraints of created order. Yet following the birth, death, and resurrection of Jesus Christ, all such promises must be seen to be at odds—even if unwittingly and inadvertently—with the divine purpose.

The redemption of creation, furthermore, is not simply a kind of generalized notion that might include the possibility of substitution, as if the original creation is to be replaced on warranty, as it were. Rather, the gospel of Jesus Christ declares the redemption of this particular creation, of the very places where we live, and of the very people that live in these places, including ourselves! In an essay contrasting the biblical vision with the modern loss of a sense of place, O'Donovan writes:

> The flesh of Jesus was particular as was no other flesh, and not simply as one among many instances of a universal rule. A universalism that responds to God's initiative has taken its beginning from the historical fact of an elect man in an elect place. If it transcends holy places, then, it does so not by subsuming them into a universal, but by proceeding from their unique, once-for-all role to new general possibilities in the history that follows them. The elect places of history are the matrix in which meetings between God and mankind are shaped.[28]

---

[27] O'Donovan, *Resurrection and Moral Order*, 14 (emphases in original).

[28] Joan Lockwood O'Donovan, "The Loss of a Sense of Place," in *Bonds of Imperfection: Christian Politics Past and Present*, by Oliver O'Donovan and Joan Lockwood O'Donovan (Grand Rapids: Eerdmans, 2004), 318.

Finally, and in respect to human being-in-the-world, the incarnation of Jesus Christ is nothing if not a colossal endorsement of embodied human being, of the very walking, talking, eating, sleeping, working, loving way of being-in-the-world that we presently and ordinarily enjoy. And although we are to be clothed in immortality at the resurrection of the dead, we will even then be recognizably embodied. After all, the resurrected Christ walked with, talked with, ate with, and was handled by his astonished disciples. "Look at my hands and my feet," he said to them. "It is I myself! Touch me and see; a ghost does not have flesh and bones, as you see I have" (Lk 24:39). True, the Christian tradition has from time to time lost sight of the significance of human embodiment, but in the face of the *dis*-embodying bent of modern technology, there is clearly an urgent need now to remember it.

Perhaps there is no better way to begin to do this than by remembering the remarkable arguments that Irenaeus marshaled against the second-century Gnostics. Over and against the Gnostic denigration of materiality, including that of the human body, Irenaeus underscored the tremendous significance that the Christian religion attached to the embodied earthiness of humanity. For the Christian religion teaches that, far from being something that separates us from God, our bodies both bear and reveal the divine glory. "By the hands of the Father," Irenaeus argued in *Against Heresies*, "that is, by the Son and the Holy Spirit, man, and not [merely] a part of man, was made in the likeness of God."[29] We bear the divine image, in other words, not simply in that we are rational and articulate, and not even in that we are personal and relational, but rather that we are all of these things and exercise all of these faculties and possibilities in bodies.

Far from limiting the human spirit, the body is the very means by which and in which our spirits are enabled, by the indwelling of the Holy Spirit, to love God and neighbor and to do our necessary work in the world. The suffering that now characterizes the human condition, Irenaeus stressed, is not therefore a consequence of the soul's being somehow imprisoned within a material body, but rather is due to the fact that our bodies are presently bruised and afflicted by sin and Satan.[30] Yet we are to be made whole again

---

[29]Irenaeus, *Against Heresies* 2.6.1.
[30]Dennis Minns, *Irenaeus: An Introduction* (Edinburgh: T&T Clark, 2010), 155.

in Christ, who has himself both become one of us and who has determined to remain one with us as an embodied human being. What sense would it have made, Irenaeus queried, for God to have redeemed his magnificent earthly creation only to abandon the bodies of the very creatures who were to represent him within it? No, the kingdom of God announced by Jesus Christ is not simply to consist in a spiritual union of souls with God but is rather to be an earthly kingdom, an actual place where the promises made to Abraham and the patriarchs will at last be completely fulfilled.[31] As Dennis Minns writes, representing Irenaeus's thought:

> Precisely because human beings have bodies, and will only be human beings so long as they do have bodies, it is absurd for them to aspire to any kind of monistic union with God as Spirit. Bodies, even glorified bodies, impose certain limitations on the Spirit conjoined to them. They need earth to walk on, food to eat and space to move in. Nor will they be able to avoid the bodies of other human beings walking on the earth and moving about in its space. Human society is a direct and inescapable consequence of our being embodied. So long as we are embodied, we will never be able to avoid it or do without it.[32]

Yet all of this, Irenaeus went on to stress, will take time and so will require patience. It will also require us to fully embrace our condition as embodied creatures. For until we accept the fact that we are not our own to fashion and refashion according the devices and desires of our own hearts but rather belong to the one who has created us, and unless we therefore embrace our created limitations, we can never be drawn close to the glory of the uncreated God for whom all things are possible. Indeed, Irenaeus reasoned, until we abandon all hope of making something of ourselves and by ourselves, and unless we allow the creative grace of God to work in and upon us, we cannot be made, as was intended from before the very beginning, after the pattern of Christ.

Speaking Christianly into our modern and largely secular technological milieu needs to begin, O'Donovan has written, echoing Irenaeus's insights, "with a moment of self-restraint, a patience that is prepared not to grasp":

> Israel learned that the holy, the coincidence of the social and cosmic, must in the end be waited for. . . . The Christian conception of the "secularity" of

---

[31]Irenaeus, *Adv. Haer.* 5.30.4; 5.32.1-2 (Minns, *Irenaeus*, 142).
[32]Minns, *Irenaeus*, 144.

political society arose directly out of this Jewish wrestling with unfulfilled promise. Refusing, on the one hand, to give up what it knew of God, itself, and the world, accepting, on the other, that what it knew was incomplete and demanded validation, Israel understood itself and its knowledge and love of God as a contradiction to be endured in hope. "Secularity" is irreducibly an eschatological notion; it requires an eschatological faith to sustain it, a belief in a disclosure that is "not yet" but is absolutely presupposed as the inner meaning of what we know already.[33]

If through impatience we allow this "not yet" to become "never," O'Donovan continues, we say something wholly incompatible with biblical Christianity. "The virtue that undergirds all secular politics," he concludes, "is an expectant patience. . . . An unbelieving society has forgotten how to be secular."[34]

Consummation

If the Christian religion proclaims that the coming kingdom of God is to be a physical kingdom with fully embodied human subjects, so it also announces that this embodiment will brilliantly transcend our present circumstances. Indeed, the redemption of creation in Christ promises to be far more than simply a return to the *status quo ante*. It promises a fullness and a superabundance that we can at present only begin to imagine. It promises, nevertheless, a superabundance that is material as well as spiritual and recognizable in terms of our present experience of created order. Irenaeus sought to capture something of the coming extravagance in a wonderful and justly famous passage. "The days will come," he wrote,

> in which vines shall grow, each having ten thousand branches, and in each branch ten thousand twigs, and in each true twig ten thousand shoots, and in each one of the shoots ten thousand clusters, and on every one of the clusters ten thousand grapes, and every grape when pressed will give five and twenty metretes of wine. And when any one of the saints shall lay hold of a cluster, another shall cry out, "I am a better cluster, take me; bless the Lord through me."[35]

---

[33]Oliver O'Donovan, *Common Objects of Love: Moral Reflection and the Shaping of Community* (Grand Rapids: Eerdmans, 2002), 41-42.

[34]Ibid.

[35]Irenaeus, *Against Heresies* 5.33.3, in *Ante-Nicene Fathers: Translations of the Fathers down to A.D. 325*, vol. 1, *The Apostolic Fathers, Justin Martyr, Irenaeus*, ed. Alexander Roberts and James Donaldson, rev. A. Cleveland Coxe (Grand Rapids: Eerdmans, 1979), 563.

Of course, at present we can only begin to catch the dimmest glimpse into that which, as the apostle Paul writes, "no eye has seen, what no ear has heard, and what no human mind has conceived" (1 Cor 2:9). Yet, again and on the basis of the apostolic encounter with the risen Christ, we can confidently say that the coming kingdom of God will be material as well as spiritual and that our participation in it will be as recognizably embodied human beings who will even then live and thrive within the precincts of time and space and within the created limitations of physical bodies. Indeed, the Christian hope is for a fully restored—yet still embodied—existence within a fully restored—yet still natural—world in which we will live and move and work while delighting in the living God. As N. T. Wright recently argued, the modern church desperately needs to remember that the Christian proclamation of the incarnation is not merely rhetorical and/or metaphorical. Neither does it represent a kind of category mistake, as both Platonists and gnostics, both ancient and modern, have imagined—as if to say that God could not possibly have deigned to stoop so low as to have become an actual human being.[36] Rather, the incarnation signals, as was foreshadowed in the Old Testament by the temple as that place where God chose to dwell on earth, the divine intention to restore this world, this earthly place, the place that includes each and every one of us.[37]

The incarnation, in short, both is the center and signals the fulfillment of the long-term plan of the good and wise Creator.[38] In this connection, Wright reminds his modern readers that there is little room for doubt as to what the apostle Paul meant when he announced "the redemption of our bodies" (Rom 8:23). He was proclaiming the redemption and fulfillment of embodied human existence.[39] "What Paul is asking us to imagine," Wright insists, "is that there will be a new mode of physicality, which stands in relation to our present body as our present body does to a ghost. It will be as much more real, more firmed up, more bodily, than our present body as our present body is more substantial, more touchable, than a disembodied spirit."[40]

---

[36]N. T. Wright, *Surprised by Hope: Rethinking Heaven, the Resurrection, and the Mission of the Church* (New York: HarperOne, 2008).

[37]Ibid., 96.

[38]Ibid.

[39]Ibid., 147.

[40]Ibid., 154.

C. S. Lewis sought to describe this "new mode of physicality" at the end of his Narnia stories. The world the characters enter following the final battle is like the old world they have left; it is recognizable in its form and reality yet it is somehow even more real, more substantial, and more deeply meaningful. Reacting to the others' surprise that this should be so, one of the older characters mutters under his breath: "It's all in Plato, all in Plato."[41] In a wonderful image, Lewis ends the story with his characters running joyfully into their new, more real world, crying, "Further up and further in!" "And we can most truly say," the narrator concludes, "that they all lived happily ever after":

> But for them it was only the beginning of the real story. All their life in this world and all their adventures in Narnia had only been the cover and the title page: now at last they were beginning Chapter One of the Great Story which no one on earth has read: which goes on forever: in which every chapter is better than the one before.[42]

The Affirmation of Ordinary Life-in-the-World

Of course, the grand Christian narrative is more nuanced than we have been able to indicate in our short précis, and each of the major elements of its story has been subjected to different and occasionally conflicting interpretations and understandings over the course of the church's long history. Lewis's mention of Plato brings one of the more specifically modern disagreements to mind. As we discussed in our last chapter, Christian theology had from very early on made good use of Platonic realism, and the assimilation of Aristotle's immanent realism within Christian theology had been the crowning intellectual achievement of the Christian Middle Ages. Yet this rich theological and philosophical synthesis, the *via antiqua*, as it came to be called toward the end of the medieval period, showed the unfortunate tendency—evident even in the concluding sentences of Lewis's Narnia epic cited above—to minimize the importance of ordinary lived reality. Because everything that really mattered lay beyond death in that realm compared to which ordinary lived reality is but a dim reflection and a shadow, there was

---

[41]C. S. Lewis, *The Last Battle* (Glasgow: Collins, 1956), 160.
[42]Ibid., 172.

obviously no great need to pay too close attention to the things of this present world. Indeed, it was believed to be spiritually perilous to do so. To gain the goods of this world at the cost of one's life in the world to come was tragic and the height of foolishness, Christianly understood.

Yet this reticence to pay too close attention to the things of this present world was bound to elicit a reaction, and the reaction eventually came in the form of what we have come to know as the modern world. Indeed, modernity may be defined in its rejection of the medieval (now read: "backward") neglect of ordinary lived reality. As Charles Taylor has observed in this connection, the cultural revolution of the early modern period was precipitated precisely by frustration with the supposedly loftier activities of speculating about ideals like "the Good" and what constituted "the Good Life." Early-modern thinkers urged us to direct our attention instead toward the goodness of ordinary life-in-the-world.[43] "Do not believe," the Florentine humanist Coluccio Salutati wrote to a friend at the close of the fourteenth century,

> that to flee the crowd, to avoid the sight of beautiful things, to shut oneself up in a cloister, is the way to perfection. In fleeing from the world you may topple down from heaven to earth, whereas I, remaining among earthly things, shall be able to lift my heart securely to heaven. In striving and working, in caring for your family, your friends, your city, which comprises all, you cannot but follow the right way to please God.[44]

From the perspective of this nascent modern outlook, theological and philosophical contemplation of such things as "the Good Life" smacked of pride and self-absorption. Early-modern thinkers also pointed out that the *via antiqua* could only really be pursued by an elite minority, that it was inherently inegalitarian, and that it required authoritarian political systems to maintain it. Instead, they argued that our principal concerns, as Taylor put it, "must [now] be our dealings with others, in justice and benevolence and on a level of equality: to increase life, relieve suffering and foster prosperity. To

---

[43]Charles Taylor, "Spirituality of Life—and Its Shadow: Today's Spiritual Innovators Turn Away from the Transcendent," *Compass: A Jesuit Journal*, May-June 1996, www.questia.com/read/1G1-30261359/spirituality-of-life-and-its-shadow-today-s-spiritual; see also Charles Taylor, *A Secular Age* (Cambridge, MA: Harvard University Press, 2007).

[44]Salutati, cited in Alan D. Gilbert, *The Making of Post-Christian Britain: A History of the Secularization of Modern Society* (London: Longman, 1980), 21.

lead one's ordinary life rightly in this way is open to everyone."[45] This affirmation of the value of the ordinary has continued to be a major component of the modern ethical outlook.

While the intellectual origins of the distinctively modern turn toward the ordinary may be traced back to late-medieval scholastic skirmishes between nominalists and realists, for practical purposes the turn received its decisive stimulus in Protestant piety. As Taylor went on to note, it was Protestant piety that "exalted practical *agapē* and was polemically directed against the pride, elitism and self-absorption of those who believed in 'higher' activities or spiritualities":

> Consider the Reformers' attack on the supposedly "higher" vocations of the monastic life, which were meant to mark out elite paths of superior dedication. In the Reformers' view, they were in fact deviations into pride and self-delusion. The really holy life for the Christian was within ordinary life itself, living in work and household in a Christian and worshipful manner.[46]

In "On the Babylonian Captivity of the Church" (1520) Luther had contended, for example, that the works of monks and priests "however numerous, sacred, and arduous they may be, these works, in God's sight, are in no way superior to the works of a farmer labouring in the field, or of a woman looking after her home."[47] All human works, Luther had insisted, are before God gauged by faith alone.

The Protestant estimation of the spiritual import of ordinary life, something that both fostered and was subsequently amplified by the careful empiricism and experimental disposition of early-modern science, unleashed an enormous amount of creative practical energy within early-modern European society and culture. Indeed, and as we saw in our last chapter, an essentially Christian—or at least Protestant—spirit lies behind the enthusiasm with which the mechanical world picture was applied to modern problems and ambitions. This distinctive spirit (though now secularized) is even now discernable beneath the surface of contemporary technological development.

---

[45]Taylor, "Spirituality of Life."

[46]Ibid.

[47]Luther, cited in John Dillenberger, ed., *Martin Luther: Selections from His Writings* (New York: Anchor, 1962), 311.

Ironically, of course, the same rhetoric originally used by the Protestant Reformers against Roman Catholic monks and priests would eventually come to be deployed by secularists and unbelievers in the eighteenth and nineteenth centuries against the Christian religion as such. The problem with Christianity, from this secular point of view, is that it scorns the real, sensual, earthly, and ordinary human good for some putatively "higher" end, the pursuit of which can only impede the realization of this-worldly progress. As, for example, Ludwig Feuerbach explained,

> The work of the self-conscious reason in relation to religion is simply to destroy an illusion—an illusion, however, which is by no means indifferent, but which, on the contrary, is profoundly injurious in its effect on mankind; which deprives man as well of the power of real life as of the genuine sense of truth and virtue.[48]

Or, as Freud repeated several generations later,

> Men cannot remain children forever. . . . Need I confess to you that the sole purpose of my book is to point out the necessity for this forward step? . . . By withdrawing their expectations from the other world and concentrating all their liberated energies into their life on earth, they will probably succeed in achieving a state of things in which life will become tolerable for everyone and civilization no longer oppressive to anyone.[49]

Or as a contributor to *The New York Times* put it in a piece entitled "Lofty Ideas That May Be Losing Altitude,"

> The idea that there is such a thing as eternal life and that it is in most important ways more important than this life. . . . has generally caused immensely greater misery than it has helped the world. First of all, it has completely devalued, for many of those who believe in it, their present life. Second, it has made many of those people who believe in it live in constant fear and guilt for what is going to happen to them afterward. And it has prevented them from doing anything to get rid of that fear and guilt by acting right because, if anything, they will get their just deserts later on.[50]

---

[48]Ludwig Feuerbach, *The Essence of Christianity*, trans. George Elliot (New York: Harper Torchbooks, 1957), 274.
[49]Sigmund Freud, *The Future of an Illusion*, trans. James Strachey (New York: W. W. Norton, 1961), 49-50.
[50]Janny Scott, ed., "Lofty Ideas That May Be Losing Altitude," *New York Times*, November 1, 1997.

Just as Protestant piety originally affirmed, and continues to affirm, the religious significance of ordinary life over and against the elitism implied by Roman Catholic practice, self-consciously secular humanists have poured contempt upon religion-in-general since the eighteenth century in the name of the affirmation of ordinary life.

We are living today in the aftermath of what Taylor termed a "double revolution," the first sparked by the Protestant affirmation of ordinary life, and a second consisting in a kind of secularized transposition of Protestant priorities since the eighteenth century.[51] The irony in this secondary development is that when traditional religions are either rejected outright or else demythologized such as to be rendered supposedly relevant within this world, our actual experience of ordinary lived reality comes to feel flat, trivial, anti-heroic, meaningless, and to many, intolerable. This, in turn, has unleashed a kind of contempt for ordinary life in the anti-life, anti-humanist stance of Nietzsche and others.[52] As Taylor observed in this connection, Nietzsche rebelled both philosophically and practically "against the idea that our highest goal is to preserve and increase life and to prevent suffering, and he rejected the egalitarianism underlying the affirmation of ordinary life."[53]

The secularization of the Protestant affirmation of ordinary life since the eighteenth century has led a number of mostly Roman Catholic theologians to contend that for the church to speak to characteristically modern problems—including that of the dehumanizing thrust of modern technological development—it will need to recover the "sacramental ontology" that lay at the heart of Christian Platonism.[54] And it would appear to be the case that when life in this world ceases to be understood in terms of eternal life, the affirmation of ordinary lived reality does indeed become somewhat insipid and less than inspiring, which has probably left modern people vulnerable to varied and sundry false promises of transcendence, including that

---

[51]Taylor, "Spirituality of Life."
[52]Ibid.
[53]Ibid.
[54]See, for example, Henri de Lubac, *Catholicism: Christ and the Common Destiny of Man*, trans. Lancelot C. Sheppard and Elizabeth Englund (San Francisco: Ignatius, 1988); see also Hans Boersma, *Nouvelle Theologie and Sacramental Ontology: A Return to Mystery* (New York: Oxford University Press, 2009); also Hans Boersma, *Heavenly Participation: The Weaving of a Sacramental Tapestry* (Grand Rapids: Eerdmans, 2011).

of neo-gnostic transhumanism. Still, the repristination of Christian Platonism is neither the only nor the most obvious solution to the problem of modern secularity. From the Christian point of view, the affirmation of ordinary life need only be grounded in Christian hope, hope that, as the Nicene Creed states, "looks for the resurrection of the body, and the life of the world to come." As theologian Emil Brunner commented,

> The New Testament says we are living in a wicked world. Therefore to live as a Christian in the State means above all to hope for the new world which lies beyond history—beyond a history which always was and will be the history of States—for that world where death and killing, force, coercion, and even law will cease, where the only "power" which will then be valid is the power of love. It is the *meditatio vitae futurae* [meditation on the life to come] which makes it possible for the Christian to do his difficult duty in this political world without becoming hard; and it is this which prevents him from lapsing into irresponsibility out of the fear of becoming hard. Both his joyful readiness for service and his sanity in service spring from this hope.[55]

The paradox of Christian existence, Brunner observed elsewhere, is that its oft-maligned "otherworldliness"—its confident hope for the future—has always been and remains the wellspring of its strength for the renewal and preservation of human cultures.[56]

Blessed Burdens

Modern technological development, we have said, is often legitimated by its promise to disburden our lives, to make them "better," more "fulfilling," "easier," more "convenient," "cheaper," and so forth. Modern technologies promise to free us, as much as possible, from the drudgery, pain, and suffering that plagues ordinary human existence. As Charles Taylor observes, modern secular understanding views all of these problems as "exogenous" to human life, which is to say that it understands them either as temporary social and/or historical systemic anomalies—often traced to late capitalism and/or patriarchy—or else it sees them as somatic and/or psychological

---

[55]Emil Brunner, *The Divine Imperative: A Study of Christian Ethics*, trans. Olive Wyon (Philadelphia: Westminster, 1947), 482.

[56]Emil Brunner, *Christianity and Civilization: The Gifford Lectures: Part II, Specific Problems* (New York: Charles Scribner's Sons, 1949), 141.

pathologies that are, or at least ought to be, amenable to therapy.[57] From the modern secular perspective, Taylor notes, what we stand to learn from our suffering is precisely nothing.[58] It is no small wonder, then, that postmodern people fall so easily for the promises of modern technology, even when many of these promises have not hitherto actually been kept.

Yet as we have also seen, falling for technology's promises of ease and convenience often actually impoverishes us far beyond the money we spend on technological gadgetry. For in seeking to evade the burdens of ordinary life, we accidentally short-circuit the natural resistance that appears to be requisite for deep learning and retention.[59] We rob ourselves of those experiences that, while often difficult and painful, make our lives real. Along this line, and expanding on the work of Albert Borgmann, theologian Richard Gaillardetz writes:

> Many technological advances rob human existence of "friction." Yet the experience of friction is one of the essential qualities that gives ordinary human existence texture—it is what makes our existence "real." It is precisely the "rough fit," the "mixed bag" of so many human interactions that brings freshness and vitality to our lives.[60]

Even more significantly, Borgmann notes that in seeking to evade the burdens of ordinary existence we can inadvertently render ourselves insensitive to the work of God in our lives; for the thorns and thistles of existence are very often the occasions of God's grace.[61] Indeed, precisely because our burdens often cause us to cry out for God, they can become translucent events enabling us to recognize and experience the work of the Holy Spirit in our lives.

From a Christian point of view, then, the afflictions and burdens that so often accompany ordinary embodied existence are actually a privileged place for encountering the living God. To be relieved of all burdens, as modern technology so often promises, and to be allowed to escape from the

---

[57]Taylor, *Secular Age*, 618.
[58]Ibid., 621.
[59]See Nicholas Carr, "The Degeneration Effect," chap. 4 in *The Glass Cage: Automation and Us* (New York: W. W. Norton, 2014), 65-85.
[60]Richard R. Gaillardetz, *Transforming Our Days: Spirituality, Community, and Liturgy in a Technological Culture* (New York: Crossroad, 2000), 42.
[61]Albert Borgmann, "Pointless Perfection and Blessed Burdens," *Crux* 47, no. 4 (2011): 28.

merely ordinary into the brilliant convenience of "hyperreality," as Borgmann puts it,[62] would be to obviate the possibility of these small, perhaps, but nevertheless invaluable epiphanies. Trying to find our lives in this way would actually be to lose them.

Along very similar lines, though in a different context, Martin Luther contended that it is not simply that God can be known in the time of suffering but rather that it is precisely in our suffering that God has determined to make himself known. Luther's radical construal of the potential of the frictions, afflictions, and suffering of ordinary life to bring us into contact with God, outlined in the Heidelberg Disputation of 1518, has come to be known as Luther's "Theology of the Cross." In a book detailing Luther's counterintuitive thesis, theologian Alister McGrath has gone so far as to contend that it was neither Luther's understanding of justification, nor his emphasis upon the final authority of the Holy Scriptures, that constituted his real theological breakthrough. Rather, it was his understanding of the redemptive possibilities of human suffering.[63]

Basically, Luther reasoned that sinful human beings cannot know God directly. To attempt to do so—as, for example, by way of the intelligent contemplation of created order, perhaps aided by classical philosophy—is to entertain what Luther called a "Theology of Glory," a theology that fails to take a full account of the devastating effects of human sin. On the contrary, he argued, in our sinfulness we are not ready to and, indeed, not yet capable of beholding the living God directly. For our sakes, therefore, God remains hidden. The place he hides, furthermore, is in our suffering—as revealed to us in the suffering of Christ on the cross. Thus, "far from regarding suffering or evil as a nonsensical intrusion into the world," McGrath writes, summarizing Luther's position, "the 'theologian of the cross' regards such suffering as his most precious treasure, for revealed and yet hidden in precisely such sufferings is none other than the living God, working out the salvation of those whom he loves."[64]

---

[62]See Albert Borgmann, "Matter and Science in an Age of Science and Technology," *Crux* 47, no. 4 (2011): 36-45.
[63]See Alister E. McGrath, *Luther's Theology of the Cross: Martin Luther's Theological Breakthrough* (Oxford: Basil Blackwell, 1985).
[64]Ibid., 151.

Luther's theology of the cross runs entirely counter to modern secular thought's dismissal of suffering as worthless, and it provides us with an entirely new vantage point from which to evaluate the "promise of technology." For while modern secular thought tends to assume that the end of disburdening human life justifies any and all technological means, the theology of the cross suggests that to attempt to eliminate, by means of modern technology, the suffering that attends the human condition, and to attempt even to overcome death, betrays an implicit theology of glory. As such it exacerbates the hubris that lies at the root of the human condition in sin—the hubris that is the cause of our suffering in the first place. Of course, we must be careful how we say this, for the love of neighbor most certainly entails healing the sick, visiting the prisoner, and caring for widows and orphans. Yet it is nevertheless in our hardships and in our sufferings where the living God often chooses to meet us. In our sufferings we very often experience God's grace, and in difficulties we very often learn humility and patience. Modern technology's promise to relieve us of suffering and hardship must, therefore, be prayerfully considered. Such promises, as the Genesis narrative indicates, may conceal insidious temptations.

Conclusion

We began this chapter lamenting Christian acquiescence in the face of the disembodying and, therefore, dehumanizing trajectory of modern technological development. We went on to suggest that what appears to account for the failure of North American Christians to protest the mechanical modern outlook, as well as our failure to resist the distinctively modern habit of objectifying and enframing the natural world, is that we have fallen out of the habit of reflecting upon and living out of the implications of basic Christian convictions. While the Christian church always stands in need of remembering its theology, the need today is particularly acute, given how rapidly automatic machine technology is trending away from ordinary embodied human life. As Erazim Kohak has noted,

> The problem could be said to be one of the uses of technology. Beneath that, however, the deeper problem is one of forgetting, of the covering-up of the moral sense of the cosmos and of human life therein beneath a layer of artifacts

and constructs. Philosophy [Theology!] has many tasks, yet in our age the task of . . . uncovering the forgotten sense of the cosmos and of our lives therein, may be one of the most urgent.[65]

Yet here it must be said that the point of remembering our theology is not simply to sound the alarm about the trajectory of automatic machine technology. Rather, the reason for remembering what the Christian religion tells us about just where we are, who we are, and the kind of work we have been given to do in this world is so that we might begin to use our energies and our considerable talents—including our remarkable capacity for scientific knowing and technological making—to enable created things, including the creatures who have been created after God's own image and likeness, to become more fully themselves. We have touched on a few of the ways that we might begin thinking about doing this. In our final chapter we will continue exploring the implications of core Christian convictions vis-à-vis modern technological development. Having now discovered a place from which "to comprehend technique from beyond its own dynamism," in other words, let's see where this can take us.

---

[65]Erazim Kohak, *The Embers and the Stars: A Philosophical Inquiry into the Moral Sense of Nature* (Chicago: University of Chicago Press, 1984), 26.

CHAPTER FIVE

WHAT ON EARTH SHALL WE DO?

*If the human economy is to be fitted into the natural economy in
such a way that both may thrive, the human economy must be built
to proper scale. . . . A proper human sound, we may say, is one that
allows other sounds to be heard. A properly scaled human economy
or technology allows a diversity of other creatures to thrive.*

WENDELL BERRY, "GETTING ALONG WITH NATURE" (1982)

IT IS OFTEN SAID that modern technology is not the problem; rather the
problem is what we do with it. This is true as far as it goes. Technology per
se—even in its distinctively modern form—is not the real problem. What we
do with technology, however, is shaped by who in the world we think we are
and by the kind of world we believe ourselves to be living in. Here we appear
to have certain problems. Our extraordinary capacity for technological
making—a capacity encompassing skills and abilities that have in recent times
arisen in conjunction with the mechanical world picture of modern science—
has made it difficult for us to imagine any other way of seeing and understanding our world. This technological facility has also made it difficult for us
to imagine any other way, beyond that of engineering, of construing our own
purposes in a world that has been thus enframed. Again, as Grant put it so
incisively, "The very substance of our existing which has made us the leaders
in technique stands as a barrier to any thinking which might be able to comprehend technique from beyond its own dynamism."[1] Combined with the

---

[1]George Grant, "In Defense of North America," in *Technology and Empire: Perspectives on North
America* (Toronto: Anansi, 1969), 40.

monetary interests that attach to technological innovation, our failure to gauge modern technological development from outside of our distinctively modern and mechanical mindset has left us insensitive to the technological diminution of ordinary embodied human being. It has also left us insensible to the beauty, wonder, and otherness of created nature. This is the peculiar modern conundrum we have been trying to understand and explicate.

The preceding chapter sought to surmount the barrier by rehearsing the crucial story elements of the Christian narrative for the sake of remembering what kind of world we actually live in as well as how the human task in this world ought to be construed. We live in a creation that has been ordered to life and fruition, a created order in which the human task—though temporarily marred by sin and death—is even now to enable created things, including people, to become most fully themselves. Facilitating this fruition requires us—at the least—to understand the compass and shape of our ordinary embodied circumstances as part of a divinely ordained order to which God has signaled his absolute commitment in the resurrection of Jesus Christ from the dead. The Christian proclamation of the incarnation and resurrection heralds God's astonishing commitment to ordinary embodied human existence, among other things.

In this final chapter I want to rehearse a number of basic Christian convictions about the world and about our place in order to try to help us discipline and give better direction to modern technological development. I say *begin* here because, so far as I am aware, no one is certain just how to do this. In truth, there probably isn't a single "Christian" way to do it. Yet from our previous discussion of the mechanical world picture, it is safe to say that any reform and/or redirection of modern technological development is going to need to begin with a change of mind. We're going to need to begin to imagine what the implications of biblical religion are for the development and use of modern technology and then to begin to live out of these implications. Just how each one of us decides to do this will be as varied and as diverse as creation itself. The purpose of this final chapter is to propose one way of thinking about the modern technological milieu and to suggest a number of questions we might ask ourselves and each other in respect to our use of modern technologies.

Our situation today is analogous to that of those New Testament crowds who suddenly realized that they were complicit in cultural systems that were at odds with God's revealed purposes. Having heard that "every tree that does not produce good fruit will be cut down and thrown into the fire," an anxious crowd asked John the Baptist, "What should we do then?" John replied that each one should do whatever was within their power to actually do. The one who had two shirts was to share with the one who had none, tax collectors and soldiers were to refrain from abusing their power, and so on (Lk 3:9-14). In a similar fashion, having learned that they had just crucified the Messiah, an anxious crowd asked Peter and the apostles, "Brothers, what shall we do?" "Repent and be baptized," Peter responded, "in the name of Jesus Christ. . . . And you will receive the Holy Spirit. . . . Save yourselves from this corrupt generation!" (Acts 2:38-40). When asking what we should do in light of modern technology's diminution of ordinary embodied human being, we must similarly repent, turn to the Lord, and begin to try to do those things that are actually within our power to do.

But what, in respect to modern technology, are we to repent of? Two things chiefly: we must—as always—repent of human hubris, of the prideful desire after autonomy, as if it were simply up to us to define ourselves for and by ourselves. We must also turn away from the enframing of the world that so clearly reflects human hubris. The principal precept of Christian discipleship is that we are not our own. From the perspective of the Christian faith, furthermore, created nature cannot be construed as stuff that stands by waiting for us to decide whether we will place a value on it—and, assuming that we decide that it is of some value to us, how we might put it to use. Created nature, including human nature, is not ours to construe in this way, neither greedily in terms of monetary profit, nor even responsibly in terms of the human values of resource management and/or sustainability. As Psalm 24:1 declares, "The earth is the LORD's, and everything in it, the world, and all who live in it." Our task, therefore, is primarily one of stewardship.

Acknowledging our need for repentance, what things are actually within our power to do with respect to the technological milieu in which we are presently enmeshed? Two insights from Jacques Ellul are helpful here. The first is his insistence that we are not, as Christians, called to fix the world.

No, Ellul reminds his readers, repairing our damaged world is God's responsibility. Acknowledging this should not lead to an attitude of resignation,
however. While we must surrender the belief that we can save the world, we
must nevertheless refuse to abet the disintegrating tendencies currently at
work in it. "We must not say to ourselves," Ellul chides his readers, "'We can't
do anything about it!' To talk like this is to play into the hands of the Prince
of this world."[2]

Christian responsibility to the modern world, Ellul goes on to stress, is
twofold: on the one hand, it is simply to seek to make the modern world
endurable so that the gospel stands some chance of being heard. The
Christian, he writes:

> must plunge into social and political problems in order to have an influence
> on the world, not in the hope of making a paradise, but simply in order to
> make it tolerable—not in order to diminish the opposition between this world
> and the Kingdom of God, but simply in order to modify the opposition be-
> tween the disorder of this world and the order of preservation that God wills
> for it—not in order to "bring in" the Kingdom of God, but in order that the
> gospel may be proclaimed, that all men may really hear the good news of
> salvation, through the death and resurrection of Christ.[3]

For the good news to be proclaimed and heard, Ellul stresses, there must
be freedom. Over and against the many apparent necessities and deter-
minisms of modern technological culture, Christians must model the possibilities of grace and freedom. It is fundamental, Ellul emphasizes, "that
people again be able to make their own decisions and be bearers of
freedom."[4] Christians are called to bring "as much free play as possible" into
government, the economy, and cultural life.[5] In this connection, Ellul insists that if we look hard enough we will always find some room to move
freely, personally, and graciously, even within situations apparently delimited by technique.

---

[2] Jacques Ellul, *The Presence of the Kingdom*, 2nd ed. (Colorado Springs: Helmers & Howard,
1989), 35.
[3] Ibid.
[4] Jacques Ellul, *Perspectives on Our Age: Jacques Ellul Speaks on His Life and Work*, ed. William H.
Vanderburg (Concord, ON: Anansi, 1981), 110.
[5] Ibid.

Determining where this room is requires careful analysis, however. It requires us to understand where we are, sociologically speaking, and to understand what is actually going on around us. An additional insight from Ellul is helpful here. Ellul believed that modern civilization is helpfully understood in terms of three layers, which he often described on the analogy of the ocean with its waves, currents, and depths.[6] On the surface are the constantly changing events and personalities covered so assiduously by mass media as well as, Ellul lamented, by mainstream academic sociology. Beneath this raucous surface layer, however, are the forces and undercurrents that are actually driving history. These subsurface forces and processes are those with which we must contend, he stressed, if we are to bear witness to God's redemptive purposes and if we are to stand any chance of effecting positive change. Ellul's fundamental contention, of course, was that modern civilization is now almost entirely driven by the logic of La Technique, a process that we have already discussed at some length. Finally, lying at a depth beneath even world-historical forces are those fundamental givens— human and otherwise—that are so often the concerns of philosophers and theologians. One of these human givens has been—and remains—the primordial human grasping after autonomy, a depravity still clearly discernable within our technological society.

In what follows, I will use Ellul's tripartite schema to identify the kinds of questions that we might usefully raise concerning our use of modern technologies as well as concerning the direction of modern technological development. One set of questions has to do with the disembodying and dehumanizing impact of specific technologies. A second set arises from our understanding of rationalization, a process discernable beneath the surface of current events and that is, perhaps, ultimately responsible for the depersonalizing bent of modern technological development. A third set of questions arises out of a Christian understanding of the human condition in sin as applied to modern technological civilization. The answers that we give to these three sets of questions comprise the "So what?" of our analysis of modern technology.

---

[6]Andrew Goddard, *Living the Word, Resisting the World: The Life and Thought of Jacques Ellul* (Waynesboro, GA: Paternoster, 2002), 119.

At the Surface

While not wanting to overlook the many and remarkable benefits of modern machine technologies and systems, there is little doubt that modern machine technology is having an adverse impact upon us in any number of respects. Indeed, and as we have seen, there is a large and growing body of evidence to suggest that certain features of our modern technological societies are not necessarily—and perhaps not *at all*—conducive to ordinary embodied human thriving. The current trajectory of modern technological development appears, furthermore, to be away from ordinary embodied human existence, which would seem to portend only more and greater disembodiment and dehumanization.

Keeping this in mind, let's begin our consideration of what we might now do by pondering another one of Ellul's remarkable insights into the technological milieu. *Technique*, Ellul states provocatively,

> is opposed to nature. . . . It destroys, eliminates, or subordinates the natural world, and does not allow this world to restore itself or even to enter into a symbiotic relation with [*Technique*]. The two worlds obey different imperatives, different directives, and different laws which have nothing in common. Just as hydroelectric installations take waterfalls and lead them into conduits, so the technical milieu absorbs the natural. We are rapidly approaching the time when there will be no longer any natural environment at all.[7]

But isn't this grossly overstated? Isn't modern technology "modern" precisely because it makes such good use of science's finely detailed and increasingly comprehensive understanding of the natural world? Ellul would not dispute this. Rather, he observes that although we must make use of nature, our modern technological use is distinctly unnatural in the sense that it travels unfalteringly in the directions of uniformity, homogeneity, and standardization. It narrows the range of options rather than broadening it; it reduces diversity rather than expanding it. Ellul sought to capture this peculiar modern thrust in the phrase "the *one* best means."[8] The very genius of the modern technological system, he believed, lay in its thoroughgoing

---

[7]Jacques Ellul, *The Technological Society*, trans. John Wilkinson (New York: Vintage, 1964), 79.
[8]Ibid., 21 (my emphasis).

evaluation of all of the possible technical means available for any given task so as always to select the single most efficient one.[9]

The modern technological drive toward homogeneity and standardization for the sake of efficiency lies at the root of modern technology's tendency to depersonalize and disembody human life. People are idiosyncratic, their individual capacities differ, and their bodies are imperfect and inefficient. And so, Ellul concludes:

> The combination of man and technique is a happy one only if man has no responsibility. Otherwise, he is ceaselessly tempted to make unpredictable choices and is susceptible to emotional motivations which invalidate the mathematical precision of the machinery. He is also susceptible to fatigue and discouragement. All this disturbs the forward thrust of technique.[10]

From the perspective of our secular culture, the divergence of modern technology from ordinary embodied human being is troubling—but it is apparently inevitable and irreversible. Technological development signals a need for rapid human adaptation, hopefully aided by newer and even better technologies. Such would appear to be the surest route to continued economic growth, and—who knows?—perhaps technological disembodiment heralds a new stage in human evolution.

Created nature, by contrast, has been divinely ordered to diversity, to pluriformity, to extravagant fruition and superabundance, to harmony, and to beauty. Within the created order, there are many ways to thrive and a great many solutions to environmental and/or other kinds of problems. Taken as a whole, created nature is efficient—supremely so, for at the end of the day nothing is ever wasted. Yet in contrast to narrow technological efficiencies, the scope of natural efficiency is ecological and global. Obedient to God's command, created nature seeks to fill time and space with fruitfulness, though of course it is frustrated at present by human sin. Following on from this contrast between modern technology and created nature, we might say that the logic of technique ultimately betrays the will-to-power—or at least a will-to-control—while created nature is enlivened by that respect for otherness that, as we have seen, is requisite for love.

---

[9]See Darrell Fasching, *The Thought of Jacques Ellul: A Systematic Exposition*, Toronto Studies in Theology 7 (New York: Edwin Mellen, 1981), 16.

[10]Ellul, *Technological Society*, 136.

What this means is that we should be suspicious of homogeneity, of standardization, of one-size-fits-all solutions, of repetition, of selfsameness and
uniformity. Under modern conditions, sameness very often betrays the logic
of *La Technique* and suggests that diversity and particularity have been—and
are being—reduced for the sake of commercial and/or mechanical efficiencies.
Authors Wendell Berry[11] and Michael Pollan more recently have pointed this
out in respect to modern industrial agriculture and modern food culture. "The
hallmark of the industrial food chain," Pollan writes in *Omnivore's Dilemma*,
is "monoculture,"[12] that continuous cultivation of single crops in the same
places that has been made possible by modern machinery and chemical fertil-
izers and that has been driven by the demand for "fast food" at standardized
outlets.[13] Pollan's concern is that while this simplification of natural complexity
for the sake of commercial manageability appears to produce consistently superior crop yields—and beautifully consistent French fries—it also damages
soils and leaves crops increasingly vulnerable to disease. The practice, combined with any number of other modern systems, seems to have given rise to
a kind of monocultural Western diet and all of its attendant health problems.[14]
Industrial food culture, Pollan fears, has blinded us to the lines of connection
and responsibility that once united us to the foods we eat. "Specialization," he
writes, "makes it easy to forget about the filth of the coal-fired power plant that
is lighting this pristine computer screen, or the back-breaking labor it took to
pick the strawberries for my cereal, or the misery of the hog that lived and died
so that I could enjoy my bacon."[15] What most commends home cooking,
Pollan concludes by way of contrast, is that it offers a powerful corrective to
this peculiarly modern way of being-in-the-world.[16]

   This is not to say that uniformity is always or necessarily bad but only that
it can signal the triumph of modern technical rationality. We will discuss
the reasons for this below under the heading of rationalization, but suffice

---

[11]See Wendell Berry, *The Unsettling of America: Culture and Agriculture* (Berkeley, CA: Counter-
point, 1977).
[12]Michael Pollan, *Omnivore's Dilemma: A Natural History of Four Meals* (New York: Penguin, 2006), 8.
[13]Ibid., 45.
[14]Ibid. 8.
[15]Michael Pollan, *Cooked: A Natural History of Transformation* (New York: Penguin, 2014), 20.
[16]Ibid.

it here to add one final observation of Pollan's. The problem with monoculture, he writes, may be as much a problem with modern culture as it is
of agriculture:[17]

> Indeed, the monocultures of the field and the monocultures of our global
> economy nourish each other in crucial ways. The two are complexly inter-
> twined expressions of the same . . . impulse . . . to elevate the universal over
> the particular or local, the abstract over the concrete, the ideal over the real,
> the made over the natural.[18]

The modern technological subordination of natural diversity for the sake of
such things as cost-effectiveness, predictability, and convenience, in other
words, is one in which we are all implicated. Every time we opt for fast food
we play our small but crucial part within the modern monocultural system.

We must also determine to take modern technology's long-standing
promise to disburden us from the toil and drudgery of the traditional past
with several very large grains of salt. While this promise has been kept in
many respects, modern technology has also saddled us with additional and,
in certain respects, far heavier burdens that stem from disengagement from
one another, as well as disengagement from immediate physical reality.
Conceivably, these burdens stem from the fact that our technologies are not
yet good enough, and we might suppose that newer and better technologies
will successfully reconnect us with each other and with reality. Unfortunately, this is doubtful. As Borgmann has noted, whatever newer and better
technologies lie just around the corner, they will almost certainly be made
available to us as commodities to be purchased and consumed, and as such,
they will probably not demand much from us in the way of commitment,
discipline, and/or skill. New technologies will no doubt be more enter-
taining and diverting, but this, Borgmann opines, will likely only "lead to
distraction, the further scattering of our attention and the atrophy of our
capacities."[19] "It is already apparent," he laments, "that the new video technology is not used by people as the crucial aid that finally allows them to

---

[17]Michael Pollan, *The Botany of Desire: A Plant's-Eye View of the World* (New York: Penguin, 2002), 226.
[18]Ibid., 228.
[19]Albert Borgmann, *Technology and the Character of Contemporary Life: A Philosophical Inquiry*
  (Chicago: University of Chicago Press, 1984), 151.

develop into the historians, critics, musicians, sculptors, or athletes that they have always wanted to be. Rather the main consequence of this technological development appears to be the spread of pornography."[20]

There is, in short, more than enough evidence to make us doubt that newer and more technology must necessarily make our lives better. Of course—who knows?—new technologies might make our lives better in certain respects. Yet unless they somehow reconnect us to each other and to our world, they will probably only exacerbate the kinds of problems we enumerated in chapter one.

In addition to fostering diversity and looking askance at the equation of newer with better, the most important thing we must do in light of the divergence of modern technological development from ordinary embodied existence is to reaffirm the basic biblical truths that, as Dietrich Bonhoeffer stressed in his exposition of Genesis 1–3, "A human being *is* a human body,"[21] and, furthermore, that Christ's incarnation is an extraordinary endorsement of ordinary embodied human being. The implications of this endorsement extend in a great many directions. Christ's incarnation affirms the primacy of our ordinary embodied human relationships. It affirms our shared experience that it is—and will continue to be—in ordinary embodied relationships with each other and with created nature that human beings will be and will become most fully themselves.

Because our embodied relationships with one another are ordinarily consummated eye-to-eye, as it were, a Christian social ethic must prioritize face-to-face relationship. As Jewish philosopher Emmanuel Levinas emphasizes, extending a patently biblical insight, the human body is recapitulated in the human *face*. It is in the face of the other where we encounter them most fully as themselves. As Bernhard Waldenfels writes describing Levinas's position, "The otherness does not lie behind the surface of somebody we see, hear, touch and violate. It is just his or her otherness. It is the other as such and not some aspect of him or her that is condensed in the face."[22]

---

[20]Ibid., 151.

[21]Dietrich Bonhoeffer, *Creation and Fall: A Theological Exposition of Genesis 1–3*, trans. Douglas Stephen Bax, in *Dietrich Bonhoeffer Works*, vol. 3 (Minneapolis: Fortress, 1997), 77 (emphasis added).

[22]Bernhard Waldenfels, "Levinas and the Face of the Other," in *The Cambridge Companion to Levinas*, ed. Simon Critchley and Robert Bernasconi (Cambridge: Cambridge University Press, 2002), 65.

Moral consciousness, Levinas is concerned to stress, is awakened in the face-to-face encounter. It is standing face-to-face where we realize that we are responsible for each other.[23] "It is a responsibility," Levinas writes, "that contains the secret of sociality, whose total gratuitousness . . . is called 'love of one's neighbor'—that is, the very possibility of the uniqueness of the one and only. . . . It is a love without concupiscence, but as irrefrangible [that is, incapable of being broken] as death."[24]

And so we must ask, Are our technologies enhancing ordinary embodied face-to-face relations, for example, by creating and/or protecting time and space for them? Are our devices making these relations more vivid and meaningful? If they are, then we ought to use them with deep gratitude. If, for example, we find that our use of social media makes us more attentive to the needs of others, then wonderful! We should use it gratefully. Yet if we find, instead, that our technologies are—perhaps for the very reasons we have detailed—undermining our ordinary, embodied, face-to-face relations with each other, then this should give us pause. It may be that technology is interposing itself between us, making our communication less fluid and our interaction less meaningful. As Andy Crouch observes, "Technology is in its proper place when it starts great conversations. It's out of its proper place when it prevents us from talking with and listening to one another."[25] If the latter is the case, then we should seek either to reform that technology or refrain from using it. To fail to do so is necessarily to undermine the possibility of empathy and understanding. There are times, in short, when we may just need to turn our smartphones off.

But the question arises: how would we know if our technologies were impeding our relations with each other rather than facilitating them? We're so dependent upon and so entangled in the use of our devices that it is often difficult to tell. It is for just this reason that many have advocated taking regular technology "fasts," that is, times that we purposely set aside to go without our computers, smartphones, televisions, and other technological

---

[23]Emmanuel Levinas, *Entre Nous: On Thinking-of-the-Other*, trans. Michael B. Smith and Barbara Harshav (New York: Columbia University Press, 1998), 105.
[24]Ibid., 169.
[25]Andy Crouch, *The Tech-Wise Family: Everyday Steps for Putting Technology in Its Proper Place* (Grand Rapids: Baker Books, 2017), 20.

devices. It is often only after we've been forced to go without our devices for a period of time that we realize how dependent upon them we have become, as well as what we may have missed in giving so much of our time and attention to them. "Once we have made the choice to give our devices a rest," Crouch writes, "once we have gotten over the crucial, core discomfort of declaring that we will not attend to them for extended periods, every single day, week, and year—we are far more likely to live with them in restful ways the rest of the time."[26]

For similar reasons, Crouch recommends arranging our homes such that the places where we spend the most time are largely free of technological devices. Fill the heart of your home, he writes, with things "that reward creativity, relationship, and engagement."[27] The screens, gaming consoles, music players, and other devices can still be there, but just not in the emotional center of the home. This, Crouch believes, is perhaps the most crucial nudge toward trying to live a "tech-wise" life:

> to make the place where we spend the most time the place where easy every-where is hardest to find. This simple nudge, all by itself, is a powerful antidote to consumer culture, the way of life that finds satisfaction mostly in enjoying what other people have made. It's an invitation instead to creating culture—finding joy in shaping something useful or beautiful out of the raw material of the world.[28]

The incarnation and resurrection of Jesus Christ must also be understood to affirm the value and standing, not simply of embodied human being, but of created reality as a whole, ordered as it is by space, time, and what we have called place. While gnostic longings after some presumably more spiritual or otherwise better reality might, prior to the incarnation and resurrection of Jesus Christ, have been plausible, they are no longer tenable. On the contrary, Christian theologians roundly repudiated gnosticism almost from the very beginning. So we must ask, Do our technologies enhance our actual experience of ordinary reality? Do they give us time to take life in? Time to really listen? Time to think and reflect? And

---

[26]Ibid., 102.
[27]Ibid., 71.
[28]Ibid., 80.

are our technologies grounding and rooting us in places? Are they enabling us to dwell richly in those places where we are most at home? If so, then—again—we should use them thankfully and gratefully. But if instead we find that our technologies have so compressed time that it has become unlivable, or if we find that our technologies have somehow flattened what may once have been places into mere space, then—again—we should either work to reform them or leave off using them. The evaluations we make along these lines need not be vexed and/or complicated, and the decisions we make with respect to various technologies may vary from one person to the next. Yet in the current climate of technological enthusiasm, in which we are encouraged to eagerly anticipate—and to save up for—whatever is to come next, basic questions about the impact of our technologies upon ordinary lived experience are not even being raised. Christians, of all people, should be asking these questions, for just articulating them is to be salt and light in the current environment.

Several Christian thinkers who have encouraged their readers to ask basic questions in respect to their use of modern technology are Wendell Berry, Albert Borgmann, and, most recently and as we have seen, Andy Crouch. Berry's questions are intended for those of us considering "up-grading" to purportedly new and improved tools or devices. He encourages us to take a hard look at the promise of new technologies to improve the quality of our lives, asking, in effect, Will they really? Is the new tool, Berry wonders, actually cheaper than the one it replaces? Is it at least as small in scale? Is the work the new tool does clearly and demonstrably better than that which is being replaced? Does it use less energy? Is it repairable, and by a person of ordinary intelligence, provided that he or she has the necessary tools? Does the new tool replace or disrupt anything good that already exists, including family and community relationships?[29] Unless such questions can be answered satisfactorily, Berry's sensible recommendation is that we save our money.

As mentioned in the introduction, Borgmann has contended that our use of technological devices—in spite of their promises to expand the range of

---

[29]Wendell Berry, "Why I Am Not Going to Buy a Computer," in *What Are People For? Essays by Wendell Berry* (San Francisco: North Point, 1990), 170-77.

things and possibilities conveniently available to us—has frequently only resulted in the loss of skills, habits, and shared practices that once graced and oriented our lives.[30] Under the headings of "focal things" and "focal practices," Borgmann has sought to encourage his readers to ask themselves a simple question: are our technologies enabling us to develop the disciplines, habits, and skills that make for an enriched engagement with reality? If they are, then by all means we should use them. Yet if our technologies are somehow undermining our engagement with reality, then we must surely reconsider our use of them. It may be that our devices are really only functioning to distract us from those things and practices that we ought to be giving ourselves to. Our evaluation need not be dour, and the goal is not austerity. A focal thing is simply something that demands our full attention and requires skilled engagement. A musical instrument is one such thing; a simple boat is another, or a tool that requires artistry and finesse. We know that we have arrived at a focal practice, furthermore, when we can answer "no" to the following questions: Is there anywhere else I would rather be? Is there anything else I would rather be doing? Is there anyone else I would rather be with? If, finally, we find ourselves thinking, "This is something I will remember," then it is likely that we have landed on what is, for us, a focal practice. Pay attention to these things and practices, Borgmann advises. They are what make life rich. Don't allow the use of technological devices to obviate or otherwise undermine them.

Given the examples he provides—musical instruments, the fireplace hearth, meal preparation[31]—we might be led to think of Borgmann's focal things and practices solely in terms of leisure or avocational activities. Berry reminds us, however, that similar satisfactions and pleasures ought come from work as well as leisure. Indeed he suggests that it is pleasure that actually perfects good work.[32] Pay attention to this, Berry urges. If we are not deriving pleasure from our work, then something is amiss. "In the right sort of economy," Berry writes,

---

[30]Ibid., 157.

[31]Borgmann, *Technology and the Character of Contemporary Life*, 104.

[32]Berry cites the Ceylonese Tamil philosopher Ananda Coomaraswamy. Wendell Berry, "Economy and Pleasure," in *What Are People For?*, 140.

our pleasure would not be merely an addition or by-product or reward; it would be both an empowerment of our work and its indispensable measure. . . . In order to have leisure and pleasure, we have mechanized and automated and computerized work. But what does this do but divide us even more from our work and our products—and, in the process, from one another and the world?[33]

Not finding ourselves pleasurably empowered by our work, we often—almost reflexively—turn to leisure to fill the void, assuming that it is just the nature of work to be dull and unsatisfying. Berry encourages us to consider whether our problems might not also stem from the modern processes of mechanization and automation and then to ask what might be done to re-unite us with our work and our products.

Just as we ought to see the divorce of work and pleasure as a symptom of a deeper, distinctively modern problem, so Crouch contends that boredom may signal an analogous disengagement with ordinary lived reality.[34] "Boredom," Crouch writes, "is actually a crucial warning sign—as important in its own way as physical pain. It is a sign that our capacity for wonder and delight, contemplation and attention, real play and fruitful work, has been dangerously depleted."[35] Crouch observes that the English word "boredom" does not even appear in the dictionary until the middle of the nineteenth century. He wonders if boredom might not actually be a distinctly modern phenomenon.[36] Could it be the case that the very technologies that promise to relieve us of boredom today are actually making the problem worse? Could it be that these technologies leave us increasingly prone to seek empty distractions?[37] Crouch goes on to observe that modern media and screen technologies quite regularly over-stimulate us by capturing—with sex, vi-olent action, vivid colors, and noise—what we discussed in chapter one as our natural "orienting response." It should come as no great surprise to learn that this constant over-stimulation has left us ever more insensitive to the more subtle and delicate contours of ordinary reality, in effect making the real world appear increasingly lifeless and dull. The solution to this problem,

---

[33]Ibid.
[34]Crouch, *Tech-Wise*, 139-53.
[35]Ibid., 146.
[36]Ibid., 139.
[37]Ibid., 140-41.

Crouch contends, is to turn our screens off as often as possible, as well as to spend time re-learning how to appreciate the marvel of the ordinary.[38] As Crouch's daughter, Amy, observes in the introduction to *The Tech-Wise Family*, "No multitude of glowing rectangles will ever be able to replace a single bumblebee."[39]

Discerning the difference between technological diversion and those things and practices that can actually enrich our lives will require us to distinguish between genuine wealth and mere affluence, however. Wealth, Borgmann explains, is whatever truly enriches our lives. Affluence, on the other hand—typically measured in monetary terms—simply enables us to purchase and consume commodities. The two are not unrelated, of course, for engagement with focal things and participation in focal practices—which is to say, activities that increase wealth—often requires commodities to be purchased and consumed. The modern conceit, however, is to assume that whatever increases affluence must necessarily increase wealth. This equation must be openly rebutted. In this connection, Borgmann warns his readers to expect to encounter resistance, for the reform of modern technology is likely to diminish affluence even as it increases wealth.[40] In other words, while the questions we need to ask each other with respect to our use of modern technology are relatively clear-cut, the costs of actually changing the way we live with our devices are not only real but also potentially high. As we stressed in our second chapter, large and powerful monetary interests attach to the device paradigm.

Still, as Erazim Kohak stresses when reflecting on the impact of technology upon modern life in *The Embers and the Stars*, it is pointless to try to have more than we can actually love. "For without love," Kohak writes,

> the claim to having becomes void. Loveless having, possessing in the purest sense, remains illegitimate, a theft. . . . Claims of entitlement are beside the point. The basic point, strange-sounding in a world of artifacts, is of a different order—that *things need to be loved*, used, and cared for.[41]

---

[38]Ibid., 149.
[39]Ibid., 12.
[40]Ibid., 239-40.
[41]Erazim Kohak, *The Embers and the Stars: A Philosophical Inquiry into the Moral Sense of Nature* (Chicago: University of Chicago Press, 1984), 108 (emphasis in original).

When it comes to modern automatic machine technology, then, we face something of a paradox. Crouch provides a nice summary of it: modern technology, he writes, "is a brilliant, praiseworthy expression of human creativity and cultivation of the world. But it is at best neutral in actually forming human beings who can create and cultivate as we were meant to."[42] Modern technology, Crouch continues, can—as in medical and/or communications technologies—serve and even save human lives, but it does little to "form human beings in the things that make them worth serving and saving." While modern technology is a brilliant reflection of human capacity, he concludes, it actually does very little to form—and may well impede the formation of—human capacities.[43]

In summarizing this surface-layer assessment of the impact that modern technology appears to be having upon us, several things have hopefully become clear. The first is how wary we need to be of techniques and technologies that promise to deliver happiness and fulfillment by way of a kind of escape from embodied limitations. *No!* we must reply. From the point of view of the Christian religion, ordinary embodied existence has been declared good—even very good—and it is our destiny to remain fully and magnificently enfleshed within a created order that has been expressly prepared by Christ to be our home.

While humanly developed techniques and technologies may be therapeutically beneficial in temporarily restoring our bodies to health, furthermore, these techniques and technologies become particularly problematic whenever they promise to do more than this. Human ingenuity, as remarkable as it is, cannot overcome the consequences of human sin. We must be deeply suspicious of techniques and technologies that promise, even if only implicitly, to resolve the root difficulties attending the human condition. As clever and as sophisticated as it may be, modern machine technology remains a human work and therefore remains subject to the thorns and thistles of unintended consequences, frustration, and futility. Indeed, given its scale and scope, as well as the hubris with which it is often deployed, it is not surprising that modern technologies regularly and repeatedly

---

[42]Crouch, *Tech-Wise*, 66.
[43]Ibid.

produce unintentionally negative consequences that dwarf the problems they were developed and deployed to solve. Falling for technology's promises encourages false hope. It renders us insensitive and unresponsive to the actions of grace in our lives.

And so we must ask, Are our technologies really enabling us to become more of ourselves? Are they enhancing our embodied relationships with others and with the world? Or are our technologies simply doing more and more things *for* us, perhaps making our lives more convenient and/or more affluent but leaving us unchanged or even diminished? The answers we give to these questions will not be the same in every circumstance, and they will vary from one person to the next. Yet if we find that we are actually being diminished by our own technologies, then we should obviously leave off using them.

Beneath the Surface

Underneath the more-or-less obvious departures of modern technological development from the nature and requirements of embodied human existence, there lies the deeper process of rationalization. Understanding this process is crucial for grasping what drives modern technology forward, and is therefore crucial for thinking about Christian witness in our technological milieu. Rational conduct, we said, entails deliberately choosing means that are suitable to reliably and efficiently realizing desired ends, or purposes. Rationalization, then, refers to the submission of any social practice—say, science, or medicine, or law, or accounting, or manufacturing—to rational methods that have been expressly designed to produce results, enabling us to predictably control some aspect of reality. In contrast to their traditional counterparts, modern rationalized methods and procedures can be counted upon—or so it has come to be assumed in modern times—to yield the fruit of practical mastery, and modern societies have come to be driven, and are very largely characterized, by the development of methods, procedures, systems, and techniques that stand the best chance of yielding such things as health, prosperity, safety, security, comfort, and convenience.

The key theorist of rationalization, Max Weber, contended that modernity is all but defined by rationalized methods, approaches, systems, procedures, and techniques. Virtually all of modern life, Weber observed—from the agency of the welfare state, to the management of the economy, to conducting warfare, to curing diseases, to raising children—has by now been surrendered to rational methods and techniques of various kinds. This rationalization of modern social life is most obvious in the large-scale and systematic manipulation of nature with machinery, as in engineering proper. Yet a kind of engineering mentality has now been carried over into social engineering—in the workplace, in political life, and in psychology. It is evident even in the most intimate areas of interpersonal experience, in what one observer has termed the "engineering of the self."[44] This whole process, Weber contended, is animated by the simple but powerful belief that we can, in principle, master all things by calculation.[45]

The rationalization of modern social life is also—and perhaps supremely—evident in the ubiquitous role that money plays in modern societies. Money is not simply an instrument useful for measuring rational effectiveness. It is the end toward which modern technological development is very often directed. The modern preoccupation with money, Mumford stressed, interprets the apparently automatic and uncontrollable dynamism of the whole modern technological system.[46]

Undoubtedly, the rationalization of modern life has been extraordinarily successful, providing us with a great deal of mastery over our practical and material circumstances and yielding the secular fruits of health, prosperity, safety, security, convenience, and so on. Yet the process has also been profoundly dehumanizing—indeed, necessarily so—for a good deal of the volatility and unpredictability that must systematically be eliminated from the equations that link means to ends derive from human personality. As Ellul observed, this dehumanizing tendency must inexorably continue.[47] "Every intervention of man," Ellul noted, "however educated or used to machinery he may be, is a source of error and unpredictability."[48]

---

[44]Peter L. Berger, *Pyramids of Sacrifice: Political Ethics and Social Change* (Garden City, NY: Anchor, 1974), 20.

[45]Max Weber, "Science as a Vocation," in *From Max Weber*, ed. H. H. Gerth and C. Wright Mills (New York: Oxford University Press, 1946), 139.

[46]Lewis Mumford, *The Myth of the Machine: The Pentagon of Power* (New York: Harcourt Brace Jovanovich, 1964), 169.

[47]Ellul, *Technological Society*, 136.

[48]Ibid.

The process of rationalization has also driven specialization, privatization, and other large-scale processes that are often employed to interpret the character of contemporary social life. Ironically, the process also seems to be the chief culprit behind the oft-lamented inversion of means and ends, for rational mastery requires life to be divided into discrete problems for which rational and technical solutions can then be devised. Yet when life is divided up in this way, we lose sight of it as a whole—in effect, losing sight of the forest for the trees. As Ellul commented,

> And what is so reassuring as the rational? . . . All that we ask of people in this society must be rational. It is rational to consume more, to change immediately what is worn out, to acquire more information, to satisfy an increasing number of desires. Constant growth is rational for our economic system. We can take the ordinary actions of 99 percent of the population in a so-called advanced country and we shall find that the key to them is always rationality.[49]

That the ordinary rational actions of 99 percent of the populations of so-called advanced countries are even now exhausting the earth's natural resources, while at the same time hastening that time when people will no longer be required within modern technological systems, suggests that the process of rationalization may not actually be very rational, at least not from the standpoint of human thriving.

The process of rationalization helps to explain the recent rise of information processing and communications technologies, as well as their significance within the contemporary economic context. The growth of industry has required the development of control technologies able to make rapid, accurate, and continuous comparisons between present states and desired outcomes. Information processing and communications systems have therefore been developed to control complicated and large-scale industrial processes, systems that are now pervasive within industry and commerce. The rationalization of control technologies is almost certain to exacerbate the problem of technological unemployment. For as we have noted, networked computer systems with access to "Big Data" are now poised to

---

[49]Jacques Ellul, *The Technological Bluff*, trans. Geoffey W. Bromiley (Grand Rapids: Eerdmans, 1990), 161.

substitute preprogrammed algorithms for human judgment in any number of industrial and commercial contexts.

Rationalization also lies behind the emergence of the "attention economy."[50] For once the value of the information about ourselves and our online habits—information that we provide the internet simply by using it—was realized, it simply made good business sense to try to collect more of it. Google, Facebook, and other online "platforms" have thus become the collectors and purveyors of vast troves of personal information, designing the "user experience" in such a way as to encourage more and more of us to stay "connected" for longer and longer. The data thus collected is subsequently made available for purchase by advertisers and others interested in understanding and perhaps influencing our behavior.

The process of rationalization has, lastly, been inevitably secularizing. Technical means and calculations now play the roles once believed to be the prerogative of the gods—and play them much more predictably and reliably. Shaped by science, technology, and industrial capitalism, the modern world has quite literally become much more of an artificial world than that of our ancestors. It is more obviously a human construction, within which the themes of human choice, human responsibility, and human control have become central. It is not difficult to see how this artificiality has created a climate within which traditional religious faith has become increasingly implausible. As Alan Gilbert noted some years ago,

> The modern child is brought up, in many cases, without a feeling of dependence upon the tides or the seasons, the vagaries of weather, wind and harvest. The modern equivalents are almost exclusively matters of human contrivance: television programmes and railway timetables, institutionalized forms of entertainment, and the spectacle of an adult world dominated by the artificial rhythms of stock market, interest rates, wages policies, inflation and unemployment. The crises, failures and disasters of such an environment may, potentially, exceed those associated with what insurance brokers and religious believers still call "acts of God," yet they are so patently human failures,

---

[50]See Tim Wu, *The Attention Merchants: The Epic Scramble to Get Inside Our Heads* (New York: Vintage, 2016).

aberrations in an artificial system, that they evoke no obvious metaphysical response or supernatural explanation.[51]

This secularization of modern social life has made it difficult to question the relentless progression of rationalization on religious grounds. Yet in the absence of substantive religious criticism, technical rationality has become, as Ellul's comments above suggest, something of an end in itself. As Robert K. Merton commented in the foreword to Ellul's seminal work, *The Technological Society*, our modern technological civilization is thus characterized by the "quest for continually improved means to carelessly examined ends."[52]

Responding Christianly to the process of rationalization must therefore entail the *re*-examination of ends. What, we must ask, are genuinely human purposes? How are such purposes most judiciously attained? What are the costs associated with the modern system's provision of health, prosperity, safety, security, and convenience? Christians should be known for asking these sorts of questions, questions that reveal both the strengths and the weaknesses of modern rationalized procedures, methods, and techniques.

We are instinctively aware that there are aspects of our lives—friendship, marriage, and family—that must not be surrendered to modern rationalized methods and techniques. We intuit that these are basic human relationships that are not helpfully evaluated in terms of productivity, efficiency, and/or cost-effectiveness. The time has now come for us to expand our list of what must not be surrendered to include churches, schools, and community organizations—associations whose primary purpose is personal formation. It simply cannot make sense to eliminate persons from the process of personal formation. The modern process of rationalization must not be allowed to overwhelm these and other personal and communal endeavors. Rather, we must insist upon a conscientious adjustment of means to ends and insist that personal ends simply cannot be achieved through exclusively impersonal means.

The problem is not that modern social practices are rational per se, and neither is it that modern social institutions have been rationalized. Rather, the difficulty is that modern practices and institutions have for the most part

---

[51]Alan D. Gilbert, *Making of Post-Christian Britain: A History of the Secularization of Modern Society* (London: Longman, 1981), 64-65.
[52]Robert K. Merton, foreword to Ellul, *Technological Society*, vi.

been very narrowly rationalized to yield secular practical and material benefits for individual entities. One of the things that makes modernity unique is the fact that its central institutions—economic, political, scientific and technological, educational—have been rationalized simultaneously and synergistically to produce a variety of mundane practical and material goods and services.

This is not necessarily a bad thing. The secular benefits of comfort, convenience, safety, and general affluence are neither to be taken lightly nor for granted. Yet it must be admitted frankly and openly that modernity's technological accomplishments are not typically valued in terms of their truth, goodness, and/or beauty. They are not typically conceived in terms of the love of God and/or neighbor. And neither are they deployed to strengthen communities. Modern technological accomplishments are valued in large part for their commercial potential, and as we have noted, money has become the crucial metric for assessing the effectiveness of rational methods and the chief end to which a great many technical means are even now being adjusted. As a result, the cultural scope of the technological society has become narrow, aimed by and large at the production and consumption by individuals of an array of commodities.

The rationalization of retail sales, for example, is premised upon the price consciousness of rational consumers—that they can be counted upon to seek low prices. The essence of this is nicely captured in Walmart's advert "Save Money, Live Better." We shouldn't have to think about the slogan for too long to realize that this simply cannot be true for all of us all of the time. Where, after all, can the savings be expected to come from except from relentlessly lowering the costs—often by means of automation—of production, distribution, and sales? The unfortunate ones who find themselves replaced by machinery and machine systems are not going to "live better" even as the rest of us "save money." It is ironic that in many cases it is the very same people who are being eliminated who can most be relied upon to try to "save money" by shopping at Walmart. Living better by saving money is apparently something that individuals can seek to do, but not together and not in communities. The relentless rationalization of retail sales is reflected in the decline and decay of Main Streets across North America and the attendant rise of rationalized strip malls and "big box" retailers. What all of this means, I think, is that the end of "living better" needs now to be expanded to include all of us together. Along this line, committing ourselves to trying to thrive together might translate into considered decisions to actually "save less" by paying each other for personal attention, knowledge, and service. These sorts of decisions don't necessarily require us to repudiate of the logic of rationalization, but they do require us to broaden—and to humanize—our understanding of the ends that the larger process ought to be made to serve.

Just how and why modernity has come to be so narrowly focused on the individual and secular ends of comfort, convenience, safety, security, pleasure, and material affluence stems, at least in part, from the affirmation of ordinary life that is one of the legacies of the sixteenth-century Protestant Reformation. Loving one's neighbor as oneself, Protestants insisted, meant looking after their ordinary material circumstances. Modernity's relatively narrow focus on secular ends also relates to the emergence of the mechanical world picture that we discussed in our third chapter. Secular ends are tangible and measurable and therefore fit readily into scientific and technological equations. Most importantly, the contemporary stress upon merely secular ends owes to the rise of that peculiar modern outlook that Charles Taylor has sought to capture with the term "exclusive humanism."[53] In his magisterial study of modern secularity, A Secular Age, Taylor observes that modern humanism is distinctive. It differs from its classical and/or religious antecedents in that its notion of human flourishing no longer makes any reference to transcendence, to anything beyond the empirical here-and-now that human beings could— and, more importantly, that they *should*—reverence, love, or acknowledge.[54] To Jesus' question "What good is it for someone to gain the whole world, yet forfeit their soul?" (Mk 8:36), secular modernity, in effect, replies, What good could it possibly be to have a "soul," if this means forfeiting the present world, which, after all, is all there really is?

The new secular and exclusive humanism emerged in European intellectual circles toward the end of the eighteenth century, one offshoot of what is often called the Enlightenment. It consisted of a set of secular appetites or

---

[53]Charles Taylor, A Secular Age (Cambridge, MA: Harvard University Press, 2007), 19-21.
[54]Ibid., 245.

demands—for safety, comfort, convenience, pleasure, entertainment, and so on—demands that innovations in industry would soon be developed to supply. By now it has become difficult for modern men and women to imagine that their hopes could extend beyond this world or that their aspirations could extend past the acquisition and consumption of material goods and services. Apparently people can live quite happily on bread alone, so it has turned out, as long as they are comfortable, healthy, suitably entertained, and distracted from asking troublesome religious questions. Within such an intellectual environment people fall easily—indeed eagerly—for what we have called the promise of technology.

Speaking Christianly into our highly rationalized culture requires us to *re*-present the possibility of human meanings and purposes that transcend this merely secular horizon. What kinds of people do we aspire—*ultimately*—to become? What are those things, we must ask, that—*finally*—make a human life worth living? While the technological provision of secular benefits such as health and security—and even comfort, convenience, and pleasure—is by no means to be taken for granted, this provision cannot, Christianly speaking, be seen to exhaust human purposes. While modern systems and devices can, furthermore and as we have seen, help us to carve out space in our lives for focal things and practices, our technologies are at best neutral in forming human persons capable of communion with each other, with created nature, and with the living God. This is not the fault of our technologies, and it is not a limitation that further technological development might be expected to overcome. Rather, modern technology's rather limited usefulness in the formation of human persons is simply a reflection of its penultimate and merely secular status.

Combined with the momentum and inertia of modern commercial and technological systems that have been designed to supply—and to profit from supplying—an astonishing array of secular benefits to modern consumers, the ongoing plausibility of exclusive humanism within modern intellectual culture is undoubtedly going to make the reexamination of human ends difficult. Those who raise the sorts of questions we have been suggesting will no doubt be castigated as "Luddites" or worse. The questions themselves may well be blunted by being recast in a therapeutic mold, in which it may be conceded that having some sense of ultimate purpose is perhaps necessary for the present experience of well-being. For these reasons we may find it necessary to employ irony and other forms of indirect communication to make the basic point that, as remarkable as modern technological civilization may be, it is still almost exclusively focused on *ephemera*, things that must soon pass away.

No one was better at poking fun at the narrowness and pedestrian scope of modern aspirations than Søren Kierkegaard, who saw through the façade of modern secular civilization already at the beginning of the nineteenth century. "In fact, what is called the secular mentality," Kierkegaard wrote,

> consists simply of such men who, so to speak, mortgage themselves to the world. They use their capacities, amass money, carry on secular enterprises, calculate shrewdly, etc. perhaps make a name for themselves in history, but themselves they are not; spiritually speaking, they have no self, no self for whose sake they could venture everything, no self before God—however self-seeking they are otherwise.[55]

Not having a self before God, Kierkegaard continued, is not something that modern secular civilization cares much about. Little fuss is made about the loss of the soul. "The greatest hazard of all, losing the self," he commented, "can occur very quietly in the world, as if it were nothing at all; No other loss can occur so quietly; any other loss—an arm, a leg, five dollars, a wife, etc.— is sure to be noticed."[56] Such mischievous comments were pointedly intended to expose the narrowness of exclusive humanism.

Foundational to the question of what ends constitute genuinely human purposes is our understanding of human personhood, an understanding that has come under relentless assault in the technological context. As we have seen, the process of rationalization finds it difficult to digest the individuality, unpredictability, and idiosyncrasies of actual human persons. The process tends either to reduce persons to functionaries and/or standardized units of one kind or another, or to eliminate persons from the machinery altogether.

---

[55]Søren Kierkegaard, *The Sickness unto Death: A Christian Psychological Exposition for Upbuilding and Awakening*, ed. and trans. Edna H. Hong and Howard V. Hong (Princeton: Princeton University Press, 1983), 35.

[56]Ibid., 32-33.

A crucial aspect of Christian witness within the technological milieu must, therefore, be the defense of the personal. As Ellul commented, the Christian fight of faith is not a fight against other people, but rather it is a fight for their freedom. This, Ellul stressed, is to "undertake the one, finally indispensible liberation of the person of our times."[57]

Unfortunately, the North American church seems largely unaware of the pressing importance of defending the personal. It appears instead to be enamored with devising and deploying impersonal and technical solutions to the problems confronting religious life. Eugene Peterson lamented several years ago concerning American Christianity:

> More often than not I find my Christian brothers and sisters uncritically embracing the ways and means practiced by high-profile men and women who lead large corporations, congregations, nations, and causes, people who show us how to make money, win wars, manage people, sell products, manipulate emotions, and who then write books or give lectures telling us how we can do what they are doing.[58]

The problem, Peterson continued, is that the ways and means most commonly employed to "get things done" in the modern context are often conspicuously impersonal, consisting of techniques, procedures, programs, general guidelines, and the like. Even if these ways and means are proven to work—and many of these techniques and programs do work, even if only in the sense of mobilizing crowds of people—they inadvertently undermine the formation of persons. Thus—however unintentionally and inadvertently—they undermine the gospel of Jesus Christ.

It is true that the value and importance of individual agency—and at least apparently of personal agency—does continue to resonate even within the technological milieu. Yet, as our last chapter stressed, whatever remaining value our modern technological civilization continues to place upon the human person is almost entirely the legacy of Christian theological convictions. Under the onslaught of automatic machine technology, this legacy is evaporating rapidly. The North American church needs now—and urgently!—

---

[57]Jacques Ellul, *The New Demons* (New York: Seabury, 1975), 228.
[58]Eugene H. Peterson, *The Jesus Way: A Conversation on the Ways That Jesus Is the Way* (Grand Rapids: Eerdmans, 2007), 8.

to remember basic Christian theological convictions concerning the nature and requirements of persons, human and divine.

Here, recall the importance of time and place for personal formation. Both are basic aspects of created order, forming the framework within which life becomes livable. Together, they create the context for the realization of truly personal agency. To meddle with either time or place is to impinge quite directly upon—and possibly to undermine—the prospects of personal fruition. Yet meddling with time and place is precisely what increasingly powerful modern technologies have enabled us to do. Indeed, it is what we have expressly designed and developed any number of technologies to do.

Consider time. The desire to compress time is nothing new. In light of life's uncertainties and ultimately in the face of death, human beings have probably always bemoaned not having enough time. They have probably always sought to pack more experiences into any given time period. From this perspective, modern machinery—now augmented by digital information processing systems—may be said to possess a kind of universal appeal, for it apparently enables us to do more and more things in less and less time.

Yet today our machines are capable of vastly greater speeds than it is possible for us to grasp experientially or even, really, to conceive. Driven relentlessly by the social fact that "time is money," machine technology has been developed and deployed to compress time so that more and more monetary transactions—each one an opportunity for profit—can be fit into any given time period. So-called High-Speed Trading (HST) is perhaps the most obvious example of this use of technology: powerful computer systems, operating according to preprogrammed algorithms and networked together by fiber-optic cabling, place millions of buy and sell orders in infinitesimally small time periods, all for the sake of gaining an advantage—and thereby making money—in the marketplace.[59] The complexity of the algorithms is so great, and the speed with which the trades are executed is so fast, that it has taken forensic researchers years to unravel market crashes that have apparently been caused by HST. Researchers are still not certain—and may

---

[59]See, for example, Paul Virilio, *The Great Accelerator*, trans. Julie Rose (Cambridge, UK: Polity, 2012).

not ever understand—what actually transpired to trigger a number of sudden and unexpectedly sharp market downturns. HST is just one of the more spectacular and recent examples of the modern technological condensation of time. Ever since the original Industrial Revolution, the majority of complaints decrying the dehumanization of the modern workplace have been linked with what was felt to be a kind of inhuman compression of time.

We should be deeply skeptical of the frequent equation of faster with better. The "better" in this equation often only accrues to those selling the devices and/or offering, for a price, to provide the services. It should come as no great surprise to learn that those of us who purchase the devices and/or subscribe to the services rather routinely—albeit unwittingly and for the most part inadvertently—impoverish some aspect of our lives. Time-saving and time-serving are very often flip sides of the same coin.

Unfortunately, modern technology also interferes with place. Indeed, in the interests of efficiency, predictability, and planning, modern technology inexorably strives to rationalize and standardize all localities—all particular places—and to turn them into merely extended "spaces." An airport in Guangzhou is largely indistinguishable from an airport in Buffalo. A shopping mall in Seattle looks very much the same as a shopping mall in Frankfurt. A McDonald's franchise in San Francisco is, for all intents and purposes, identical to its counterpart in Cape Town. Managers and engineers employing modern machine technology reduce concrete places into commercial and/or other kinds of "spaces" for the sake of efficiency, predictability, planning, and profit. This reduction is almost universally experienced by ordinary human beings as a kind of loss, which is an important clue to the threat that modern technological systems pose to human flourishing. Although we may be able to survive for a time in these "spaces," we can only really thrive in places.

We must also, therefore, be deeply skeptical of the equation of uniformity with ease and/or convenience, particularly when this requires places to be converted into mere spaces. To be sure, it is somewhat comforting that McDonald's restaurants are the same in Moscow as they are in Montana. Yet a good deal of this comfort simply owes to the fact that the homogenization of commercial spaces renders them predictable, which reduces the stress that we are so often under because of the technological compression of time. We might ask ourselves, Where will we actually live once all places have been reduced to spaces? Imagine trying to live for any length of time inside a shopping mall—a rationalized retail system that has been expressly and carefully designed to move merchandise. Although we could, perhaps, survive there for a time, we could not make—nor would we be allowed to try to make—any kind of real life there.

In sum, we need to see that modern machine technology's relentless compression of time as well as its progressive obliteration of place must render the possibility of genuinely personal action moot. Bereft of any particular place within which to move, and deprived of the time necessary for careful and prayerful deliberation, it must become impossible for human beings to act freely, responsibly, and meaningfully.

True, the physical and cultural particularities of places resist and frequently thwart technological standardization. Yet the ideal of an entirely and utterly rationalized space has become at least virtually possible by means of that vast web of networked computer systems known as the Internet, a virtual location typically labeled cyberspace. Polls tell us that more and more people are spending more and more of their time "there,"[60] which is troubling because cyberspace is not and cannot ever be a real place. As Borgmann observes, "Cyberspace is not a region within physical space and at definite distances from other regions in physical space. It's a realm of its own that, once entered, is distanceless."[61] As such, he continues, it must inevitably disappoint us, for it cannot really support life. Virtual reality is simply too far removed from *real* reality, from created order. It doesn't provide enough actual resistance to make lasting human growth and development possible. To say this is not to minimize the utility of the Internet but simply to acknowledge that real people require actual places in which to live and that the virtual space made possible by computer networks simply cannot be made to substitute for actual places. In this connection, and as we have seen, empirical sociological and

---

[60]The most recent figures I could find were collected by the Barna Group and included in Andy Crouch, *The Tech-Wise Family*. North American parents report that children, for example, spend on average five hours on electronic devices on typical weekdays (p. 108).

[61]Albert Borgmann, "Grace and Cyberspace," *Crux* 47, no. 4 (2011): 8.

psychological studies have found that social media—to cite one example—distort and may actually impede real human communication.

Yet, cyberspace conveys the illusion of vitality, and a great many people spend a good deal of time in cyberspace precisely because of all of the things that seem to be possible there. Borgmann attributes the seductive appeal of the Internet to its carefully crafted presentation of what he calls "hyperreality."[62] Hyperreality is a kind of pseudo-reality that boasts several advantages over reality per se: *brilliance*, highlighting the desirable elements of experience and minimizing and/or excluding the less-desirable elements; *intensity*, insofar as it is experienced as "better than real" and comprises more of what we might hope to get from experience; and *pliability*, in the sense that it is interactive and at least apparently subject to our management and control.[63] Vast resources and the efforts of a great many very talented people are even now marshaled and deployed to design the internet "experience" such that we might spend more of our time—and more money—in cyberspace. As Borgmann goes on to lament,

> The actual world is in danger of declining to a resource of cyberspace, the realm from which we draw the materials and the energy to maintain and expand cyberspace, the place where those actual persons live whose homely and prosaic lives provide the raw material for hyperreal representations on the Internet.[64]

This transmutation of the real into the virtual cannot be good news for real human persons. The neglect of the actual for the sake of the merely possible is, as we have seen, a recipe for psychosis. That said, vast sums of so-called smart money are even today funding firms dedicated to producing technologies that promise to more effectively despoil the actual for the sake of the profits that are waiting to be made in cyberspace.

To conclude this middle level of analysis: a process of rationalization operates under the surface of modern technological societies, a process that has a great

---

[62]Albert Borgmann, *Crossing the Postmodern Divide* (Chicago: University of Chicago Press, 1992), 82-97.

[63]Ibid.

[64]Albert Borgmann, "Matter and Science in an Age of Science and Technology," *Crux* 47, no. 4 (2011): 40-41.

deal to do with the trending of modern technological development away from ordinary embodied human persons. The process is one in which rational producers, employing the insights of modern science and utilizing increasingly sophisticated control technologies, engineer and manufacture commodities designed to be sold and expended by rational consumers whose desires and preferences have been shaped in such a way as to be fleeting and inexhaustible. The process strives after maximal efficiency and endless, quantifiable growth. It is no accident that this essentially impersonal process has exacerbated—and will surely continue to exacerbate—the problems of disembodiment and dehumanization within modern societies, for ordinary embodied human persons are very often the source of error and unpredictability within rationalized systems. Resisting the process of rationalization requires us to understand it, to ask what sorts of things constitute genuinely human purposes, and to defend the possibility of genuinely personal agency within the technological milieu.

At Depth

As we have seen, the distinctively modern drive to rationalize more and more aspects of life has arisen out of a mechanical world picture that some trace all the way back to classical Greek metaphysics. Several episodes in the genealogy of this mechanistic outlook were detailed in chapter three. It is important to stress, however, that the modern habit of enframing the world as lifeless stuff also reflects one of two possible answers that can be given to a timeless metaphysical question. In trying to place his finger on the metaphysical temper of modern technological culture, Kohak frames the question as follows:

> Shall we conceive of the world around us and of ourselves in it as *personal*, a meaningful whole, honoring its order as continuous with the moral law of our own being and its being as continuous with ours, bearing its goodness—or shall we conceive of it and treat it, together with ourselves, as *impersonal*, a chance aggregate of matter propelled by a blind force and exhibiting at most the ontologically random lawlike regularities of a causal order?[65]

Is the person the root metaphor of thought and practice, or is the root merely matter in motion? With that question answered, Kohak insists, all else follows.

---

[65]Kohak, *Embers and the Stars*, 124-25.

Kohak's articulation of this question is both frank and profound, and surely all else has followed. That our technological civilization has, in effect, decided the question in favor of matter in motion is indicated by any number of things. In the first instance, it is explicit in modern scientific understanding. Contrasting modern science's view of the world with that of Genesis, for example, Kass observes:

> According to [modern] science, the universe is not a cosmos, is not an integral and finite whole. The splendid heavenly bodies are not ensouled or divine, but are, like us, limited and perishable. Moreover, nature is not teleological or purposive. Beneath the surface of things, everything obeys the laws of nature, but in the world of our experience all is flux, and chance and necessity rule, indifferent to our concerns or even to our survival. On the earth, not only beings but also species come and (mostly) go, while the vast emptiness of space preserves its icy silence. For these reasons, especially, nature as understood by science has nothing to teach us about human good: the descriptive laws of nature do not issue in normative Natural Law or Natural Right. In this respect above all, nature by herself does not provide us a true home.[66]

On such a bleak account, human consciousness is simply a kind of happy (or unhappy) cosmic accident that we will probably never fully understand. However human existence is explained, we have apparently been left by the blind forces of nature to define ourselves over and against a universe that appears to be indifferent, and is perhaps inimical, to our survival. Thankfully, so the modern scientific account continues, we are clever, flexible, and adaptable organisms with a robust instinct for survival. What choice do we have, then, but to survive as best we can by adapting to our changing circumstances? To the extent that technological ingenuity enables us to carve out space for ourselves amidst the icy flux of things, and to the extent that scientific knowing and technological making appear to provide us with engaging aspirations, then so be it. Let our technologies take us where they will, and let us hope for the best. That, at any rate, is how the modern argument tends to run, an argument that necessarily follows from the initial assumption that the universe is explicable without remainder in terms of matter in motion.

---

[66]Leon R. Kass, *The Beginning of Wisdom: Reading Genesis* (New York: Free Press, 2003), 5-6.

The metaphysical decision to interpret reality exclusively in terms of matter in motion also underlies the moral and social constructivism that characterizes what is commonly called the postmodern condition, that view of the world that assumes that all human meanings—metaphysical, moral, cultural, and the like—are socially constructed over and against a backdrop of meaninglessness and ultimately of chaos. In the contemporary context, scientism, technological nihilism, and moral and social constructivism tend to reinforce each other. As Kohak notes, only in a context in which nature had somehow been completely eclipsed by human artifice could we ever have gotten to the point of imagining the natural world to be meaningless. We have, he concludes, become victims of self-forgetting, so bedazzled by our technological accomplishments that we have become blind to the otherwise manifest moral sense of nature.[67]

That we have, as a civilization, decided that the root metaphor of thought and practice as impersonal is indicated, finally, in the unrestrained consumption that characterizes modern technological society. Unremitting consumption betrays a desperate attempt to fill with possessions the deep emptiness that stems from believing ourselves to be alone in an essentially meaningless universe. Like the Slough of Despond in Bunyan's *Pilgrim's Progress*, this existential void is bottomless. The consumption of goods and services can never finally fill it. This doesn't keep postmodern men and women from continuing to try, however. After all, what else is there to do? It is in this connection that Kohak sheds challenging light on the ancient biblical command "Thou shalt not covet." The command, he writes,

> is not an injunction against the rightful striving of all beings whose being is projected into temporality. It is an urgent warning against turning the world from the place of our dwelling into an object of possession, rendered dead and soulless by greed. Of all the commandments governing the relationship of finite beings to each other, it is, perhaps, the most basic. No force is more destructive than greed, no drive more elemental. Greed is not an extension of need, since a need can be satisfied. [Greed] is the desperate attempt to fill with

---

[67]Kohak, *Embers and the Stars*, 24.

possessions the emptiness which humans create when they ignore the first four commandments, turning their world into a meaningless wasteland in which they are utterly alone.[68]

Kohak is surely correct. Greed is undeniably one of the basic forces with which we will have to contend in seeking to restrain and to give discipline to modern technological development. Yet something more must be said, for as the apostle Paul reminds us, greed is idolatry (Eph 5:5), and it ultimately discloses fear. Those who are most greedy are often those who are most afraid. Perhaps the most basic obstacle to restraining and disciplining modern technological development, therefore, is fear: the fear that neither the world of chance aggregates of matter propelled by blind forces, nor that of money and increasingly impersonal machinery, can ever really provide a home for us; the fear that we are alone in a meaningless universe; the fear, ultimately, of death.

Yet the question arises: So what is the problem for which such an impersonal interpretation of reality has seemed to offer a solution? This question can be answered in terms of the history of ideas, much as we have done in our discussion of the intellectual genealogy of the mechanical world picture. Viewing the world as a vast and elaborate machine seemed to hold out the hope that if we could only discover how natural mechanisms worked, we could make them work for us. Yet it is crucial to stress that the question must also be answered theologically, for the mechanical world picture is not simply the result of a unique line of historical and intellectual development. Neither is it merely traceable to a kind of metaphysical mistake. Christianly understood, the impersonal worldview that lies at the root of modern technological development must ultimately be traced to human sin, for it betrays hubris, that rebellious bid for autonomous self-definition that is the essence of sin. A mechanical world, after all, is a world amenable to rationalized methods, procedures, and techniques. It is a world that is open, at least in principle, to human management. It is a world in which autonomous human mastery is not merely conceivable but is for all intents and purposes compulsory. A mechanical world, in short, is one in which exclusive

---

[68]Ibid., 79.

humanism makes a great deal of sense, and in which the determinations of good and evil are ours to decide and ours to impose upon a reality that no longer has any meaning or significance apart from whatever clever uses we can devise for it.

In New Testament terms, the mechanical outlook fostered by modern machine technology is just another iteration of the sinful human construction at enmity with God that the apostle John calls "the world" at the beginning of his Gospel, a "world" that while having been made by Christ nevertheless refuses to recognize his authority (Jn 1:10). As Ellul puts it,

> The world today, the world according to Scripture, is not God's good creation of life, love and freedom in which humans in communion with God mediate his presence in the world. It is the counter-creation of rebellious human beings and the powers bearing the marks of its rejection of the Creator in death, closure/unity, Eros, and necessity. It is the lost world of defiance and opposition to God.[69]

Christians, Ellul argues, are called to be present at precisely those points of maximum tension between a world that is, in effect, bent on suicide and God's redemptive purposes for it. "Our concern," he writes, "should be to place ourselves at the very point where this suicidal desire is most active, in the actual form it adopts, and to see how God's will of preservation can act in this given situation."[70] Our central contention is that the modern world's suicidal desire is most active—and indeed most obvious—in modern technology's drive to diminish ordinary embodied human being.

What this means is that Christians must be prepared to defend, not simply human persons and the possibilities of genuinely humane and personal action, but the deeply personal quality of reality itself. How to go about making such a defense is one of the more difficult challenges we face, for modern science's impersonal reading of nature has been tremendously successful and has proven very useful. C. S. Lewis anguished over this problem at the end of what he believed was one of his most important essays, *The Abolition of Man*. Is it possible, Lewis asked, to imagine a new natural philosophy that doesn't

---

[69]Ellul, cited in Goddard, *Living the Word*, 86; see also Ellul, *Presence*, 19.
[70]Ellul, *Presence*, 19.

confuse the "natural objects" produced by its analysis and abstraction with reality itself? He responded:

> I hardly know what I am asking for. . . . The regenerate science which I have in mind would not do even to minerals and vegetables what modern science threatens to do to man himself. When it explained it would not explain away. When it spoke of the parts it would remember the whole. When studying the It it would not lose what Martin Buber calls the Thou-situation.[71]

The difficulty of even beginning to think along such lines is not simply that our imaginations are enmeshed in the mechanical world picture of modern science but also that the modern engineering mentality has produced—and no doubt will continue to produce—all sorts of powerful and at least apparently useful technologies. Again, it is our scientific and technical prowess that most prevents us from putting technology into the proper perspective.

Still we must try to surmount this barrier. In this connection, Kohak stresses trying to recover what he calls the "moral sense of nature." "I think only a person wholly blinded and deafened, rendered insensitive by the glare and the blare of his own devices," he wrote, "could write off that primordial awareness of the human's integral place in the cosmos as mere poetic imagination or as 'merely subjective.'"[72] No, Kohak continues, the most basic reality that confronts us in nature is that it is God's world, not ours. "The heavens declare His glory," Kohak acknowledges, "the creatures of the forest obey His law, the human dweller gives thanks for His grace."[73]

Kohak's comments recall the theology of creation outlined in chapter four. The creation we said is deeply good and has been ordered by God to extravagant fruition. Christ's incarnation, furthermore, signals that God remains utterly and absolutely committed to his creation. "Fear not!" our message must therefore be; the world is decidedly not a chance aggregate of matter propelled by a blind force, and it is not, therefore, indifferent to human interests. On the contrary, created nature is precisely where we belong. It has been painstakingly and deliberately crafted by an all-powerful and, most importantly, beneficent and loving creator to provide just the

---

[71]C. S. Lewis, *The Abolition of Man* (Glasgow: Collins, 1943), 47.
[72]Kohak, *Embers and the Stars*, 6.
[73]Ibid., 182.

kind of place where we can—as embodied human beings—flourish and thrive. True, God's temple garden has been seriously, though temporarily, marred by human sin. Things at present are not as they should be. Yet human sin has neither frustrated the divine purpose nor nullified the original human vocation vis-à-vis created nature. Our task is still to take care of the creatures—including each other—on behalf and for the sake of the one who created them, a task for which we have been wonderfully and marvelously suited, not least by way of intellectual suppleness and practical ingenuity. Just as the first human pair was commissioned to care for created things, enabling those "others" to come most fully into their own being, so we are still called to this good work. Will this work be vexed by the thorns and thistles of unintended consequences? Will it be frustrated by sin and death? Of course, for the time being it will. Yet is it work that remains worth doing? Of course it is.

Yet we cannot fulfill our God-given commission if we remain insensitive to the contours of created existence, both physical and moral. Neither can we successfully carry out our charge without taking real care to respect the otherness of things. We certainly cannot remain true to our calling if we remain convinced that it is our prerogative simply to impose our will upon things. As O'Donovan writes,

> Man's monarchy over nature can be healthy only if he recognizes it as something itself given in the nature of things, and therefore limited by the nature of things. For if it were true that he imposed his rule upon nature from without, then there would be no limit to it. It would have been from the beginning a crude struggle to stamp an inert and formless nature with the insignia of his will. Such has been the philosophy bred by a scientism liberated from the discipline of Christian metaphysics.[74]

O'Donovan's comments are ironic, of course, for modern scientism has not so much been liberated from the discipline of Christian metaphysics as it has rejected the limits to human creativity and to human willing that are implied in the Christian understanding of the nature of things. Unless these limits are once again acknowledged, there can be no hope of recovering—

---

[74]Oliver O'Donovan, *Resurrection and Moral Order: An Outline for Evangelical Ethics* (Grand Rapids: Eerdmans, 1986), 52.

even if only haltingly and partially—the human vocation vis-à-vis created nature, and no hope of reforming modern technology.

Whatever hope we have for reforming modern technology must also be tempered by patience and restraint: patience, in the sense that we must not collapse the "not yet" of the coming kingdom of God into the "already" of present possibilities; restraint, in the sense that we must continue to acknowledge and respect the distinctive shape of created nature. As Irenaeus once stressed, to create is an aspect of God's special goodness, whereas being created belongs to the goodness of human nature. To find that we exist as particular embodied human creatures within the confines of our particular times and places is not a limitation that must somehow be overcome. Rather, it is something for which we ought to be deeply grateful. Irenaeus stressed that it is precisely in striving to acknowledge that it has been given to us to be ourselves and to inhabit our particular circumstances that we open ourselves up to God's creative presence in our lives. We do not make God, Irenaeus reminded his readers, but rather God makes us, and he has both promised and is most certainly able to bring us to perfection in his own good time. Wait then, Irenaeus entreated his readers, for the hand of the master, who makes everything at the proper time for all them that are created.[75]

This does not mean that we should not strive to improve our circumstances. Neither does it rule out the wonderful creativity that God has obviously bestowed upon human beings. But it does rule out seeking—impatiently and by means of technology—to redeem ourselves now. It rules out seeking—impatiently and by means of technology—to escape from our present circumstances and/or from the perceived confines of created nature. For just as impatience may well have aroused the original sin, so impatience must continue to drive us away from the divine purpose.

Our technological society seems to oscillate between effusing over the apparently limitless possibilities of modern technology and despairing over the dehumanizing consequences of technological development thus far. This odd vacillation suggests that modern men and women have grown increasingly impatient—not simply with their own perceived limitations and/or the

---

[75]Irenaeus, cited in J. Patout Burns, *Theological Anthropology*, Sources of Early Christian Thought, ed. William G. Rusch (Philadelphia: Fortress, 1981), 27.

apparent confines of created nature but ultimately with God. "God is a long time in coming," the modern argument has tended to run, "so we had better save ourselves." Yet one of the distinctive features of our now *post*-modern situation is that we have grown reluctant to place hope in merely human abilities and agencies. We appear now to be poised to try turning matters over to sophisticated machinery that has been carefully programmed with complex algorithms that promise to establish technical-rational control over our circumstances, while at the same time enabling us to escape, in effect, from ourselves. To qualify for this remarkable "salvation" we are asked only to give ourselves over to the machinery and to surrender ordinary embodied existence. This may once have seemed a rather high price to pay. Happily (at least from the standpoint of *La Technique*) the attraction of the ordinary has been diminished by the extravagant promises of new technologies and the hyperreality of cyberspace. Resistance, in any event, seems futile.

No! Christians must stand—stubbornly, prophetically, and by reason of basic Christian convictions—squarely in the path of this fruitless exchange. We must insist that when the trajectory of modern technological development is away from ordinary embodied existence, it is at odds with God's purposes for his world. Although it may seem as if we have been left to our own devices to solve our own problems, we have not. God is there, and he is for us. He will surely accomplish everything that he has promised. In the meantime, we must encourage each other to be both patient and hopeful.

Of course, from the point of view of the Christian religion, the hoped-for consummation of all things will not merely entail a return to the original garden. Neither will it necessarily entail a reversal or an unmaking of all that we have made by way of technical ingenuity. On the contrary, the Christian hope for the resurrection of the body means that much of the work that has been—and is even now being—done "in the body" will, by the grace of God, find its place in the coming kingdom. The New Testament suggests that the work that will last is work that restores and enriches ordinary embodied human existence, work that quickens and awakens us to created reality, and work that in so doing better enables us to admire and love created things as we were originally intended to love them.

Remembering what the Christian religion tells us about who we are, about where we find ourselves, and about the kind of work we have been called to do within the created order should go some distance toward enabling us to properly evaluate modern technological development. It provides a perspective from which technological development can be properly assessed. Yes, the recovery of Christian understanding will require us to repent of enframing as well as of autonomous self-construction. Yet it promises to make a new way of seeing possible. While the revelation of the sons and daughters of God for which the creation now waits in eager anticipation is a future event, we can even now—on the basis of what we already know—begin, as Wendell Berry puts it so vividly, to "practice resurrection."[76]

Conclusion: Practicing Resurrection

How do we account for the extraordinary success of modern scientific knowing and technological making, which only surfaced after nature had ceased to be understood personally and anthropomorphically, or as Weber put it, after the natural world had been disenchanted? This is a difficult question, closely related to how we might recover an understanding of the natural world that doesn't reduce all things personal to merely mechanical forces. Are scientific knowledge and technological efficacy necessarily impersonal?

It may help to recall that one of the core metaphysical insights responsible for the rise of early-modern science was the realization that in order to truly know the natural world, one needed first, in effect, to listen to it. To understand nature one needed to enter into a kind of dialogue with it. According to T. F. Torrance, this insight entered into early-modern scientific understanding by way of John Calvin's hermeneutical method. "A genuine question," Torrance notes describing Calvin's method,

> is one in which you interrogate something in order to let it disclose itself to you and so reveal to you what you do not and cannot know otherwise. It is the kind of question you ask in order to learn something *new*, which you cannot know by inferring it from what you already know.[77]

---

[76]Wendell Berry, "Manifesto: The Mad Farmer Liberation Front," in *The Country of Marriage* (New York: Harcourt Brace Jovanovich, 1973).

[77]See Thomas F. Torrance, *God and Rationality* (London: Oxford University Press, 1971), 34 (emphasis in original).

Calvin's concerns had to do with the interpretation of Scripture, his point being that we need to listen carefully to the account that Scripture gives of itself if we are to understand it rightly. Yet Torrance notes that this same carefully inductive method came to be applied by seventeenth-century researchers to their understanding of the natural world.[78] The critical scientific questions thus became, What is the actual nature of this thing that we are seeking to know? and How, given its actual nature, is it therefore appropriately known?

The carefully inductive approach to the natural world taken up by early-modern science resonates with the Christian theology of creation, for love is both appreciative and receptive.[79] It acknowledges and appreciates the other as "other," and it seeks to receive its knowledge from the other, in effect, by assiduously listening to the account the other gives of itself. It strives in this way to know the other—not as they are perhaps represented within a framework or system of knowledge that we are already certain of but rather as they actually are. From a Christian point of view, it is hardly surprising that the empirical posture almost immediately began to yield the fruit of rich understanding, for it amounted to taking a kind of I-Thou posture in respect to nature.

Yet early-modern thought quickly assumed that the only reason we would want to know nature was so that we could more effectively put it to use. "Nature," Francis Bacon (ca. 1580) famously asserted, "to be commanded must be obeyed." This early move in the direction of utility led to an approach to knowing that, as Guardini has observed, "unpacks, tears apart, arranges in compartments, takes over and rules."[80] It is a way of knowing that does not enter into dialogue with nature so much as it analyzes nature for the sake of achieving power over it.[81] It is a way of knowing that betrays that manipulative relation to nature that Heidegger sought to capture in the term *enframing*.

Yet it can at least be asked whether this manipulative move was necessary. Could not Calvin's original insight, and the early-modern turn toward the

---

[78]Ibid., 33.

[79]See Jason Lepojärvi, "How Many Loves? A Reinterpretation of C. S. Lewis's *The Four Loves*," in *The Undiscovered C. S. Lewis: New Thoughts and Directions in Lewis Studies*, ed. Bruce R. Johnson and Jerry Root (Oxford: Oxford University Press, forthcoming 2017).

[80]Romano Guardini, *Letters from Lake Como: Explorations in Technology and the Human Race* (Grand Rapids: Eerdmans, 1994 [1923]), 43.

[81]Ibid., 44.

empirical, just as easily have been employed for the sake of a fundamentally different way of knowing—a way of knowing that, as Guardini also put it, "sinks into a thing and its context" in order to "live with" it?[82] Put differently, could we not have sought to know the world of nature simply in order, as Grant put it, that we might love and admire it?

Had we not been sinful people, perhaps we could have done precisely this. As it happened, the prospects for power and profit that opened up with the new approach to nature quickly overwhelmed all other considerations, setting modern civilization off down a path of using—and abusing—nature for the sake of humanly determined interests, a path we remain on today.

Having said this, nothing prevents us from seeking to explore and perhaps to model that other, earlier way of knowing our world for the sake of loving and admiring it. This offers a unique opportunity for Christian witness at the present time, for it recalls the stewardly dominion we were originally meant to have over the created order, pointing toward our eventual reconciliation with the natural world. The witness of Christian environmental organizations like *A Rocha* might be mentioned in this connection.[83] Founded by Peter and Miranda Harris in 1983, *A Rocha* employs scientific research precisely in order to conserve and restore the natural world through practical conservation projects. "Our scientific work," their Canadian website explains, "provides the strong foundation needed for informed protection and management." Good science, they have found, is crucial in "slowing or reversing the trends of habitat loss that are affecting many species."[84]

Just as the human rule within creation began with Adam's careful consideration of the various creatures for the sake of naming them, so we remain free even now to listen to created "others" in order that we might bring them more fully into being in our speech. This may seem like a small thing, but the power of our speech should not be underestimated. As the psalmist declared, "Out of the mouth of babies and infants" God has established a stronghold against his enemies (Ps 8:2 ESV).

---

[82]Ibid., 43.
[83]See A Rocha's website: www.arocha.ca.
[84]See www.arocha.ca/what-we-do/conservation.

In connection with the reexamination of our knowing and naming within the created world, it is worth recalling Heidegger's contention that in the face of danger created by modern technology, we must rediscover a more poetic way of being-in-the-world. In an essay entitled "The Origin of the Work of Art," Heidegger suggests that our language creates a kind of space—or, as he puts it, a "clearing"—within which all of the other things that lack the power of speech can come fully into being. "Language alone," Heidegger writes, "brings beings as beings into the open for the first time. Where there is no language, as in the Being of a stone, plant, and animal, there is also no openness of beings. . . . Language, in naming beings for the first time, first brings beings to word and to appearance."[85] Heidegger believed that precisely this nomination of things to appearance has been lost within our technological culture. We have continued to use language to name things, but our names for things have become abstract and are often narrowly geared toward setting things in motion and putting things to use—toward envisioning things as standing-reserve. Our names no longer allow things to come into the fullness of their own being. At the very moment we became the masters and possessors of nature by way of technological knowing and making, nature—including human nature— seemed to disappear from our field of view as something distinct from us and as having its own integrity apart from our will-to-power. Technological enframing, Heidegger believed, has concealed other ways of being-in-the-world that more effectively let creatures—including people—come into presence as truly themselves. Could we not rather seek to know the world of nature, Heidegger hinted, in order to truly give voice to it, in order to allow things to come most fully into their being? The answer, surely, is *Yes!* We can—we *must*—try to do exactly this. Indeed, we anticipate the quality of the human relation to nature, for which the creation now waits in eager anticipation, simply by noticing its brilliance and beauty, by articulating its marvelous ordering, and by celebrating our uniquely embodied presence within it before God. After all, as the nineteenth-century poet Gerard Manley Hopkins once queried, "What is earth's eye, tongue, heart else, where else but in dear and dogged man?"[86]

---

[85]Martin Heidegger, "The Origin of the Work of Art," in *Martin Heidegger: Basic Writings* (New York: HarperSanFrancisco, 1977), 198.

[86]Gerard Manley Hopkins, "Ribblesdale," in *Poems of Gerard Manley Hopkins* (London: Humphrey Milford, 1918).

Let us simply conclude, then, with one of Hopkins's wonderful poems: "As Kingfishers Catch Fire." Not only does this poem give evidence of Hopkins's profound respect for the otherness of things, a respect that seems to have necessitated poetic articulation, but the poem seems also to capture that distinctive role that we are meant to play in the scheme of things: that of delighting in and giving voice *to* the created order.

As Kingfishers Catch Fire

As kingfishers catch fire, dragonflies draw flame;
As tumbled over rim in roundy wells
Stones ring; like each tucked string tells, each hung bell's
Bow swung finds tongue to fling out broad its name;
Each mortal thing does one thing and the same:
Deals out that being indoors each one dwells;
Selves—goes itself; *myself* it speaks and spells,
Crying *What I do is me: for that I came.*

I say more: the just man justices;
Keeps grace: that keeps all his goings graces;
Acts in God's eye what in God's eye he is—
Christ—for Christ plays in ten thousand places,
Lovely in limbs, and lovely in eyes not his
To the Father through the features of men's faces.[87]

---

[87]Gerard Manley Hopkins, "As Kingfishers Catch Fire," *The Oxford Poetry Library: Gerard Manley Hopkins*, ed. Catherine Phillips (Oxford: Oxford University Press, 1995), 115.

A PERSONAL CONCLUSION

*All Christian theology has its origin in the miracle
of miracles that God became human.*

Dietrich Bonhoeffer

Hoping to grab a quick cup of hot coffee to go, my wife and I stopped into a McDonald's several weeks ago in Hope, British Columbia. Upon entering the restaurant, we were confronted by a row of McDonald's new self-serve kiosks. As the patrons lined up behind them seemed to be having trouble using them, we proceeded to the counter, where we were informed by one of the two cashiers still staffing the registers that the only coffee drinks available were iced because the automatic coffee machine was down. We returned to the road with two Diet Cokes.

I had heard that McDonald's had announced its intention to automate more of its operations, so when we arrived home I looked it up on the web. A Google search for "McDonalds" and "automation" quickly turned up a number of hits. Some announced that the stock price for McDonald's had recently hit an all-time high (June 2017) in light of a recent announcement that the firm would begin implementing the "Experience of the Future" by replacing service workers with automatic kiosk ordering and table delivery.[1] Other hits cited former CEO of McDonald's, Ed Rensi, as divulging this: "I

---

[1]Tae Kim, "McDonald's Hits All-time High as Wall Street Cheers Replacement of Cashiers with Kiosks," CNBC website, June 20, 2017, www.cnbc.com/2017/06/20/mcdonalds-hits-all-time-high -as-wall-street-cheers-replacement-of-cashiers-with-kiosks.html.

have said that robots are going to replace people in the service industry going forward, and a self-service kiosk is nothing more than automation taking over [for] people."[2] Minimum-wage legislation and new labor regulations, Rensi suggested, in combination—we must assume—with improved robotic technology had simply made the move inevitable from a business point of view.

I mention this because it's such a good illustration of some of the various forces I have described—of the collision of the lives of ordinary people with rapidly advancing technologies and basic cost-profit analyses. Over the last several hundred years we have become very, very good at developing and deploying new technologies. We have also developed an extraordinary economic system that consistently rewards profitability. Profitability has often reflected the productivity gains that have been made possible by efficiencies realized through technologically improved control systems and automation. It is a remarkable system, and it has produced—and continues even now to produce—the highest material standards of living for the largest numbers of people in human history.

Yet ours is also a system that has from the beginning had problems with ordinary embodied human persons. Ordinary people—occasionally indolent workers, ignorant and/or irrational consumers, shortsighted policy makers, myopic politicians, even as physical beings who periodically need to rest and recuperate—always seem to be interfering with the efficient functioning of things. It is no wonder that there have been pressures from the Industrial Revolution onward to phase people out of modern systems and to replace them with automatic machine systems. It is no wonder that technological development has continued to diverge away from ordinary embodied human existence. This, of course, is one of the problems we have been concerned to redress.

All of this can make it sound as if this phasing of ordinary people out of modern systems is something that is happening *to* us, as if it is all being foisted upon us by "the Man," and we are the victims of large, impersonal "processes" over which we have little or no control.

---

[2]Matthew Wisner, "Former McDonald's USA CEO: Robots to Replace People in the Service Industry Going Forward," FOXBusiness, March 1, 2017, www.foxbusiness.com/features/2017/03/01 /former-mcdonalds-usa-ceo-robots-to-replace-people-in-service-industry-going-forward.html.

But this isn't quite true. We all participate more or less actively and willingly in the larger process and in a variety of ways. We are often mesmerized by the whiz-bang promises of the new technologies: we tirelessly surf the web in search of the best deals; we consult experts in order to invest in the most profitable firms; we seek to get a jump on the future by training in specialized skills; we often seek to avoid face-to-face contact by hiding behind our screens, and so on. We—each of us—contribute in many ways to the functioning of the larger system

Here I must offer a *mea culpa!* I am just as guilty of these sorts of things as the next person, something that, oddly enough, hit home the other night as I flipped through the pages of the latest issue of *Bicycling* magazine.

I am an avid road cyclist, a MAMIL (Middle-Aged Man in Lycra) as we are sometimes affectionately (or not) labeled. And it's true that as I've gotten older I've come to value the low-impact exercise that cycling makes possible. Yet even more, I love the smooth, quiet, and rapid movement through physical space that cycling on a nice stretch of pavement makes possible. What pure joy it is zoom down a smooth bit of asphalt on a warm day and through a beautiful landscape! There's nothing quite like it. I heard someone say once that you don't really know a place until you've cycled through it, and I've been blessed to experience the truth of this observation, both in North America and in Europe.

I further confess that I love cycling gear. There are few things more gorgeous than a smartly designed and expertly welded compact titanium road frame, unless perhaps it is the precision-machined chainrings and carbon fiber crankarms of a Campagnolo Super-Record crankset. Road bikes, I would argue, are things of beauty, true *objets d'art*. They are also veritable galleries of advanced technology, much of it developed in and for aerospace applications: titanium, carbon fiber, stainless steel, aluminum and magnesium alloys, tiny parts machined to incredibly strict tolerances, electronic activators and servo-mechanisms, power meters, GPS-enabled computers, not to mention Lycra, Styrofoam, all manner of plastics, high-definition and anti-reflective optical lenses, and super-bright LED lights. And, of course, the roads we cyclists ride on are constantly being repaired and improved by increasingly complex machine systems.

Browsing through all of the latest gear in *Bicycling*, I was struck by the constant—and completely unabashed—emphasis upon "performance" in many of its advertisements as well as in a good deal of its editorial content. Not only are the performance characteristics of various bikes and bike components frequently emphasized but also how one can increase one's own physical performance, often through mental and physical exercises of various kinds, dietary supplements, proper hydration, and the proper "fueling" of one's metabolism. The subtitle of one recent story listing tech products introduced at a recent cycling tradeshow captures *Bicycling*'s general vibe: "Sure, you might not need it, but trust us—you'll really, really want it anyway."[3]

The other night I found myself wondering: Where is the line that separates improving one's physical performance (deemed good) from performance-enhancement (deemed bad)? Why do the advertisements for so many "new" bicycling products resemble those for new pharmaceuticals, that is, replete with paragraphs of fine print warning of possible side effects? Why, furthermore, did it seem like the pieces that discussed the existential pleasures of cycling seemed mostly aimed only at those just getting into the sport, while the articles aimed at more serious riders tended very often to be technical and/or technological?

Several things stand out to me. First, much of what I was reading in *Bicycling* was written from the perspective of—and actually celebrates—what we have called the mechanical world picture, in this case envisioning the human body as a machine. Second, I was struck by the pervasiveness of the modern penchant for enframing, here extending even into an increasingly popular form of leisure. Third, I realized that I am myself deeply implicated in this peculiarly modern vision because, frankly, I love all this stuff! Finally, I began to grasp just how hard it's going to be to imagine a fundamentally different way of being-in-the-world.

All of us, it seems, stand in need of a fundamental change of *mind*. We are all going to need to revisit the basic questions of who we are, how we ought to be related to one another, what kind of place the world is, and how we

---

[3]Matt Phillips, "The Coolest New Cycling Tech at Interbike," *Bicycling*, September 22, 2017, www
.bicycling.com/bikes-gear/tech/the-coolest-new-cycling-tech-at-interbike.

ought to live in it. And most of us, myself included, are starting from scratch, as it were, in revisiting these questions because of our dependence upon—and indeed fondness for—the mechanical and technological status quo.

Our larger culture, I have argued, is stuck in a posture of enframing, having no real alternative to the mechanical world picture. Its acquiescence in the face of technological disembodiment owes to a view of the world that sees nature, including human nature, as a vast and elaborate mechanism that may differ from automatic machine technology in degree but not in kind. Embodied human existence from within this peculiar worldview is, again, most often construed not as something to be nurtured and enhanced (perhaps by technology) but rather as a series of limitations to be overcome with more and better technology.

From a Christian point of view, contemporary technology's divergence away from the requirements of ordinary embodied human life indicates a serious problem. Anything that undermines, enfeebles, or otherwise diminishes ordinary embodied human existence must be seen to be at odds with the divine purposes signaled in the incarnation and resurrection of Jesus Christ. As Dietrich Bonhoeffer once commented:

> God did not become an idea, principle, a program, a universally valid belief, or a law [or, we might add, a *system*]. God became human. That means . . . Christ does not abolish human reality in favor of an idea that demands to be realized against all that is real. Christ empowers reality, affirming it as the real human being and thus the ground of all human reality.[4]

This *Christ*ian empowering of reality, this absolute affirmation of ordinary embodied human being is very, very good news for a whole variety of reasons, not least of which is that it provides the vantage point from which to evaluate and give discipline to modern technological development.

And so we must try now to *re*-member the basic shape of Christian theology, recollecting all of the ways that this theology affirms ordinary reality and ordinary embodied human life within what was—and remains— God's good creation. We must follow up our recollection with "practicing

---

[4]Dietrich Bonhoeffer, *Ethics*, ed. Wayne Whitson Floyd Jr., trans. Reinhard Krauss, Charles C. West, and Douglas W. Stott, in *Dietrich Bonhoeffer Works*, vol. 6 (Minneapolis: Fortress, 2005), 99.

resurrection" together, trying—humbly, joyfully, hopefully—to live out of the implications of our theology. It looks like we have time. The forces and processes that we have described are not going away any time soon. The issues stemming from technological disembodiment are only going to become more and more obvious, and it would speak well of us within the larger culture, I think, if we could get out in front of them. Indeed, just beginning to ask the kinds of questions we have suggested above is to be salt and light in our contemporary context. And, as the psalmist prays, "May the favor of the Lord our God rest on us; establish the work of our hands for us—yes, establish the work of our hands" (Ps 90:17).

One additional personal vignette is in order here at the conclusion of our study. My wife spent a couple of hours on the phone with Amazon.com a week or so ago trying to track an order that had apparently gotten lost. It turned out to be one of those frustrating and annoying customer service experiences where you are shunted from one automated system to another. It wasn't until she was able to speak with an actual human representative that she was able to get the problem sorted out. As she recounted the saga, she off-handedly mentioned how often it is today that you hear people say, "Finally! . . . I was able to get ahold of a *real* person." And I thought to myself, *What a fitting way to conclude this book!* For, of course, it is our destiny in Jesus Christ to be able, someday, to say just exactly this, and in the richest, deepest, and fullest possible sense. Finally we will be able to behold as well as to become real human persons! And in the meantime we can practice attending to getting ahold of each other!

ON EUCHARISTIC EMBODIMENT

*"What is Christianity? Likeness to God as
far as is possible for human nature."*

BASIL OF CAESAREA,
ON THE HUMAN CONDITION

THE MECHANICAL WORLD PICTURE and what we have called modern gnosticism are, of course, two sides of the same coin. The mechanical world picture leaves so little room for human personality that modern men and women have been all but compelled to posit another realm—conceivably a region of pure spirit or perhaps a virtual reality made possible (ironically) by advanced technology—to which we will one day escape and within which we will at last be truly ourselves, free from the confines of actual reality as well as from the interference of others. Once the stuff of science fiction, such aspirations are uttered quite seriously today in tech circles.

Unlike our third- and fourth-century forebears, those of us in the Christian church are not currently offering a great deal of resistance to the gnostic modern drift. Protestants, in particular—in both our "liberal" and "conservative" guises—have tended to succumb to a kind of discipleship that is either indistinguishable from secular "wisdom" or removed altogether from the exigencies of ordinary life-in-this-world.[1] Liberals speak of a "spirituality" derived from a "gospel" that has been so thoroughly demy-thologized that it is difficult to distinguish from ordinary well-wishing.

---

[1] See Philip J. Lee, *Against the Protestant Gnostics* (New York: Oxford University Press, 1987).

Conservatives, on the other hand, often stress the importance of being saved *from* this world but remain studiously vague as to what this salvation might actually be *for*.

Our difficulties no doubt stem from the fact that Protestants have spent so many of the last several hundred years in dialogue with modern science and technology and may even be considered responsible for the modern mechanical outlook in certain respects. Whatever the reasons, modern Protestant worship tends to reference a sacred that is outside of and largely unrelated to this world. Modern Protestant spirituality, furthermore, tends to emphasize the individual's private experience of God and encourages believers to anticipate being released from the limitations of this world altogether.

It is perhaps not surprising that modern Protestant believers have not been particularly sensitive to the threat modern automatic machine technology poses to human embodiment. Our experience of the world has been so disenchanted—or, more accurately, desacramentalized—that we tend to see little in ordinary embodied existence that discloses the life of God. A kind of mechanical frame of mind has prevented us from seeing that the entire world, including our own lives, participates in and proceeds moment by moment from the living God, the God "in whom we live and move and have our being" (Acts 17:28). We have apparently also forgotten that when "the Word became flesh and made his dwelling among us" (Jn 1:14) this both vindicated and immeasurably exalted ordinary embodied human existence.

What can we do about this? How can we *re*-absorb the implications for us that Jesus Christ was, and remains, as the Athanasian Creed states, "Perfect God; and perfect Man, of a reasonable soul and human flesh subsisting." How can we *re*-sacramentalize our experience of the world?

First, as suggested above, we need to recover the basic Christian doctrines of creation, incarnation, and resurrection, along with all their implications. These are astonishing declarations of the relation of the Creator of all things to his creation, including ourselves. As Leonard Vander Zee observes,

> God's story and the creation's story come together in Christ, making things more than bits of matter, and opening our eyes to their ultimate transfiguration. Creation, incarnation and the ultimate re-creation of the cosmos

reveal *a God for whom matter matters,* and material things open our eyes to the One who is above and beyond all things.[2]

Next, we must strive to put this theology into practice. This was discussed in chapter five, where I suggested that this labor might be helpfully conceived in terms of "practicing resurrection." While we await the general resurrection of the dead, we said, we can nevertheless try even now to anticipate—even if only in the sense of pointing to it—life in the new creation. As a kind of coda to this discussion, I would like to suggest that there is no better place to begin practicing resurrection, and no better way to celebrate human embodiment, than in celebrating the rite that has from the very beginning constituted the Christian church: that simple yet momentous act of remembering Christ's sacrifice in the sharing of the bread and wine in what the church has traditionally called the Eucharist.

What Happens in the Eucharist?

A number of profoundly important things happen in the Eucharist. Let's start by considering the most obvious: the Lord's Supper is a focal practice that brings us together in the name of Christ to prepare and eat a simple meal around a common table. It is an intrinsically embodied rite that centers in the basic bodily requirements of nourishment and sociality as well as the basic human practices of cooking and gathering around a table to eat.

The Eucharist is also the church's principal rite of *re*-membrance, of recounting who Jesus was (and is) and what he did (and does) for us. As the apostle Paul reminded the Corinthian church,

> For I received from the Lord what I also passed on to you: The Lord Jesus, on the night he was betrayed, took bread, and when he had given thanks, he broke it and said, "This is my body, which is for you; do this in remembrance of me." In the same way, after supper he took the cup, saying, "This cup is the new covenant in my blood; do this, whenever you drink it, in remembrance of me." For whenever you eat this bread and drink this cup, you proclaim the Lord's death until he comes. (1 Cor 11:23-26)

---

[2]Leonard Vander Zee, *Christ, Baptism and the Lord's Supper: Recovering the Sacraments for Evangelical Worship* (Downers Grove, IL: InterVarsity Press, 2004), 19 (emphasis added).

But why remember and proclaim the awesome significance of the Lord's death (and resurrection!) in this peculiar way? Why around a table with a simple meal of bread and wine? Why are the bread and wine consecrated with the words "This is my body" and "This is the new covenant in my blood"?

The physical and material aspects of the Lord's Supper may be said to reflect a kind of divine accommodation to our physical weakness. Our Lord commanded us to use common, ordinary, and earthly elements in our sacramental proclamation of his death and resurrection because we are common, ordinary, and earthly beings. Seen in this light, the bread and wine may be said to manifest God's gracious condescension to us. Just as the Christ "became flesh and made his dwelling among us" (Jn 1:14), so he has deigned to continue to dwell among us in the creatures of bread and wine. "The blessing of sacramental worship," Vander Zee writes,

> is the thrill and comfort of knowing that God *meets us where we are*, washing us, feeding us, quenching our thirst for grace. We not only believe it, we sense it, see it, taste it, feel it, smell it and swallow it. What my mind doubts, my mouth tastes as the Lord's goodness. When my faith falters, my fingers can touch the truth.[3]

In his teaching on the Eucharist, John Calvin emphasized this "fleshing out" of the Eucharistic remembrance. While the intellectual capacity of our souls enables us to confess that "Christ is Lord," Calvin noted, nevertheless we most fully and personally experience Christ's saving work through the bodily senses of taste, touch, smell, and sight.[4] As physical creatures, Vander Zee stresses (recalling Calvin's teaching), we need more than simply to hear the gospel. Rather, we need to touch it, to smell and taste it, just as lovers need more than the words "I love you" but also a kiss and an embrace. In short, we require more than words to bring us into a relationship with Christ. We need *things*.[5]

Yet the Eucharist doesn't merely symbolize our union with Christ in a way that we—frail, earth-bound creatures—can begin to comprehend it. It is not

---

[3]Ibid., 25 (emphasis added).

[4]See Thomas J. Davis, *This Is My Body: The Presence of Christ in Reformation Thought* (Grand Rapids: Baker Academic, 2008), 87.

[5]Vander Zee, *Christ, Baptism and the Lord's Supper*, 24.

simply, as is sometimes (mischievously) said, a kind of "divine flannelgraph."
Rather, the church has always insisted that the Eucharist actually enacts our
union with Christ; that is, that the reception of "the most precious Body and
Blood" of Christ of which we speak in the liturgy is not simply metaphorical,
but real! Calvin, for example, went on to insist that our actual union with
Christ—a union not just with his spirit but also with his body—was the essence of the Christian life.[6] In participating in the Eucharist, he insisted, we
are not simply reminded that this union will someday take place but instead
that, in having physically eaten the bread and having drunk the wine, our
union with Christ has actually—if mysteriously—been accomplished. In
this connection, Calvin simply recalls the church's ancient teaching, formu-
lated by Augustine, that the sacrament is an "effective sign," which is to say,
it confers what it signifies.[7] The significance of this for the Christian life
simply cannot be overstated. As Ian McFarland writes,

> If Christ's entire humanity has been glorified in the resurrection, and if the
> body and blood of the risen Christ are confessed as present in the consecrated
> elements of the Eucharist, then it follows that in the Eucharist we come into
> direct contact with glorified matter (viz., Christ's body and blood) in the
> world of time and space.[8]

Our participation in Christ's *real* presence in the sacramental elements is
not simply a kind of temporary accommodation. Christ does not meet us
where we are provisionally, only then to return to a purely spiritual reality,
to which we will someday also be called. Rather, just as the incarnation of
God the Son was permanent and indicated an enduring transformation of
created reality, so our union with Christ in the Eucharist signals that we are
ourselves enduringly transformed in our ingestion of the consecrated bread
and wine. Just as the incarnation signals that God has determined to indwell
created reality in such a way as to render it even more real than it was before,
it is precisely into this "more" that we are graciously introduced in the celebration of the Eucharistic feast.

---

[6]Davis, *This Is My Body*, 88.
[7]Augustine, cited by Vander Zee, *Christ, Baptism and the Lord's Supper*, 29.
[8]Ian A. McFarland, *From Nothing: A Theology of Creation* (Louisville: Westminster John Knox,
2014), 174.

Can we actually experience this "more," this newly intensified reality in the here and now? Of course! We experience it in the actions of believers, and indeed in our own actions, as we are "made new in the attitude of [our] minds" (Eph 4:23) and thus freed to do those good works that "God prepared in advance for us to do" (Eph 2:10). As the apostle Paul insists in 2 Corinthians 5:17, "If anyone is in Christ, the new creation has come: The old has gone, the new is here!" Still, although the New Testament insists that the new creation is manifest and that we should be looking for the signs of it, this does not necessarily imply a change in physical appearances, just as the elements of the Eucharist do not look any different after they have been consecrated with the words of institution. We must trust in the Lord's promise to be present to us in the consecrated bread and wine. We must walk by faith.

It is God the Holy Spirit who unites our bodies with Christ's resurrected body. For, as the Apostles' Creed affirms, the embodied Christ is now seated in "heaven" and "at the right hand of God the Father." Although he will, as the Creed continues, "come again to judge the living and the dead," we must at present be lifted up in order to be united to him. This lifting up of our own bodies to be united with Christ's resurrected body, Calvin stressed, is the mysterious work of God the Holy Spirit:

> Even though it seems unbelievable to us that Christ's flesh, separated from us by such a great distance, penetrates to us, so that it becomes our food, let us remember how far the secret power of the Holy Spirit towers above all our senses. . . . What, then, our mind does not comprehend, let faith conceive: that the Spirit truly unites things separated in space.[9]

For Calvin, therefore, the liturgical phrase "lift up your hearts" was not simply metaphorical. Rather, at the Lord's Table our hearts are—in the power of the Holy Spirit—actually lifted up to Christ in heaven and united to his glorified humanity. Again, in ingesting the consecrated bread and wine our bodies are mysteriously united to his resurrected body. As Christ's resurrected body is the "firstfruits" of the resurrection (1 Cor 15:20), furthermore,

---

[9]John Calvin, *Institutes of the Christian Religion*, ed. John T. McNeill, trans. Ford Lewis Battles, The Library of Christian Classics 21 (Philadelphia: Westminster, 1960), 2:1370.

so in the Eucharist we become—for the first time—fully embodied, fully and eternally real. Irenaeus wrote about this reality already in the second century:

> For, as the bread that comes from the earth, when it receives the invocation of God is no longer ordinary bread but the eucharist which comprises two elements, an earthly and a heavenly, so our bodies which participate in the eucharist are no longer corruptible, since they now have the hope of resurrection.[10]

Finally, it is crucial to stress that the Eucharist is a communal meal. At the Lord's Table we eat bread and drink wine together. In the modern context of rationalized and atomized individuality, it is crucial to stress that the Eucharist both commemorates and effects our incorporation into the one body, making us members not simply of Christ but also of one another (1 Cor 10:17). As Peter Leithart observes, "Because we eat together of one loaf, we are one body. . . . called to radical Christlike, self-sacrificing love, to use whatever gifts we have for the edification of the body, to live lives of forbearance, forgiveness, and peace."[11] Celebrating the Eucharist frequently, Leithart suggests, must surely help to "inoculate" the North American church against the gnostic tendencies of our technological culture.

Eucharistic Embodiment

While we await the resurrection and glorification of our bodies (cf. 1 Cor 15:12-58), the Eucharist suggests how this glorification needs now to be understood. For while the present form of things, bounded as it is by time, is necessarily "passing away" (1 Cor 7:31), this should not be taken to imply the dissolution of embodied existence. For God in Christ has not promised to make all new things, such that the Christian hope is someday to leave this world—including our bodies—entirely behind. Rather God has promised to make all things—including our bodies—new! We thus look forward to a renewal and not a replacement of created order, a renewal in which things will be "new" because God indwells them in a new way. God cannot be any nearer to his creation than he already is. What remains to be revealed, then,

---

[10]Irenaeus, *Adv. Haer.* 4.18.5, in Willy Rordorf, ed., *The Eucharist of the Early Christians*, trans. Matthew J. O'Connell (New York: Pueblo, 1978), 91.
[11]Peter J. Leithart, "The Way Things Really Ought to Be," *Westminster Theological Journal* 59 (1997): 171.

is a kind of deepened divine intimacy in which his servants will continually "see his face, and his name will be on their foreheads" (Rev 22:4). In the celebration of the Eucharistic feast, we are given a taste of this future glory in the present. We are also embedded in the new creation. As McFarland writes,

> The Eucharist draws us upward by drawing us together, binding us not only to one another but also to the bread and wine, which in their organic connection with soil, water, sun, and air implicate the whole web of creaturely relations that makes our life specifically and genuinely human.[12]

In Christ, McFarland continues, God the Creator has also become a creature of earth. In eating the bread and drinking the wine that are Christ's body and blood, we proclaim that it is precisely because God has become a creature of earth that we, the creatures of earth, have also become the children of God. When, in the twinkling of the eye and at the last trumpet, we are "changed" (1 Cor 15:52), the nature of this change will not be such as to disembody us, somehow releasing us from the limitations of space, time, and matter. Rather, we will be changed to become more fully and more gloriously embodied in a newly *re*-created order. We will then, as the apostle Paul puts it, finally be fully "clothed" with imperishability and immortality (1 Cor 15:54). McFarland cites Maximus the Confessor in this connection: "For the Word of God wills always and in all things to accomplish the mystery of his embodiment."[13] Within Eastern Orthodoxy, the Eucharist is understood in terms of our priestly vocation within created order. Alexander Schmemann writes:

> The first, the basic definition of man is that he is *the priest*. He stands in the center of the world and unifies it in his act of blessing God, of both receiving the world from God and offering it to God. . . . The world was created as the "matter," the material of one all-embracing eucharist, and man was created as the priest of this cosmic sacrament.[14]

The Eucharist is the sacrament of human embodiment, celebrating the astonishing incarnation of God the Son as an embodied human being, mysteriously uniting—in the power of God the Holy Spirit—our bodies to Christ's

---

[12]McFarland, *From Nothing*, 180.

[13]Ibid., 181.

[14]Alexander Schmemann, *For the Life of the World: Sacraments and Orthodoxy* (Crestwood, NY: St. Vladimir's Seminary Press, 1963), 15 (emphasis in original).

resurrected body and anticipating a gloriously embodied life within a re-
splendently renewed creation, to the glory of God the Father.

The Eucharist and Human Technology

The Eucharistic proclamation of "the Lord's death until he comes" should in
no way be taken to imply the denial of human technology. On the contrary,
the outward and visible signs of our union with Christ's body and blood are
precisely the remarkable cultural—and, indeed, *technological*—achievements
of bread and wine, both of which may be said to represent human ingenuity
at its best. In the Eucharist, what we eat and drink is not simply what God gives
to us but also that which we have labored, often by means of sophisticated
technology, to produce.

Yet it is also true that the technologies employed in the Eucharist celebration are those that both require and nurture our resourcefulness and skill,
that bring us together around a common table, that are a delight to our eyes,
ears, noses, and tongues, that nourish our bodies, and that fill our hearts with
joy. They are technologies that bespeak the remarkable fruitfulness of the
created order and our unique place within it, reminding us just *where* we are
as well as *who* we are. They are technologies that render our experience of
ourselves, of each other, and of created reality more vivid and more real.

The technologies that lie behind and beneath the Eucharistic feast do not
diminish us. Rather, they enable us—in and of themselves—to become more
fully ourselves. It has been given to them to play an instrumental role in that
rite that enables us to become most fully ourselves in Christ, and that is
wonderful indeed. The Eucharist is surely a foretaste of heaven, when the
kings of the earth will bring their splendor into the new Jerusalem (Rev 21:24).

For all of our concerns about modern technology and the diminution of
ordinary embodied human existence, our technologies can surely also
enable us to become more fully ourselves, more delightfully related to one
another, more thoughtfully engaged in and with created reality, and better
attuned to the voice and will of the living God. We just need to be clear that
these things are, finally, what our technologies are *for*.

AUTHOR INDEX

SUBJECT INDEX

SCRIPTURE INDEX

Finding the Textbook You Need

The IVP Academic Textbook Selector
is an online tool for instantly finding the IVP books
suitable for over 250 courses across 24 disciplines.

ivpacademic.com